AF712163

3-Digit Place Value

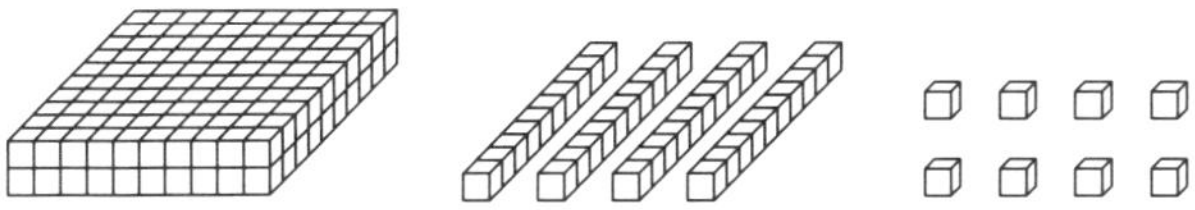

2 hundreds 4 tens 8 ones =

Hundreds	Tens	Ones
2	4	8

Write each number.

1. 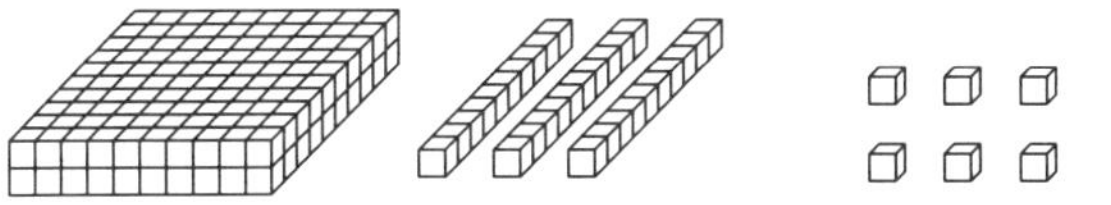

Hundreds	Tens	Ones

2. 

Hundreds	Tens	Ones

3. 5 hundreds, 6 tens, 4 ones = ___ H ___ T ___ O

4. 7 hundreds, 3 tens, 8 ones = ___ H ___ T ___ O

5. 6 hundreds, 4 tens, 9 ones = ___ H ___ T ___ O

6. 2 hundreds, 9 tens, 5 ones = ___ H ___ T ___ O

7. 9 hundreds, 4 tens, 1 one = ___ H ___ T ___ O

8. 7 hundreds, 3 tens, 6 ones = ___ H ___ T ___ O

9. 5 hundreds, 8 tens, 5 ones = ___ H ___ T ___ O

10. 6 hundreds, 1 ten, 7 ones = ___ H ___ T ___ O

11. 7 hundreds, 2 tens, 2 ones = ___ H ___ T ___ O

12. 6 hundreds, 7 tens, 7 ones = ___ H ___ T ___ O

A number can be shown in many ways.

342 with blocks

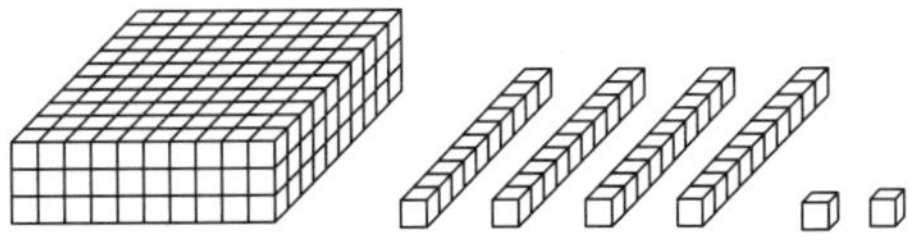

342 with pictures

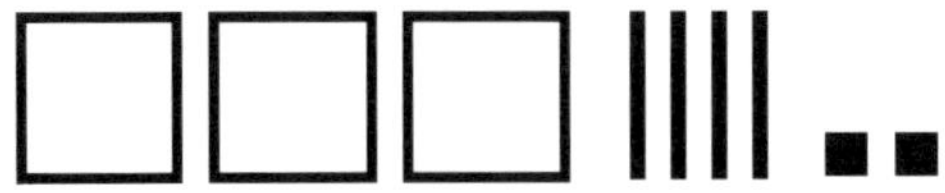

A number can be written in expanded form using place value.

342 =

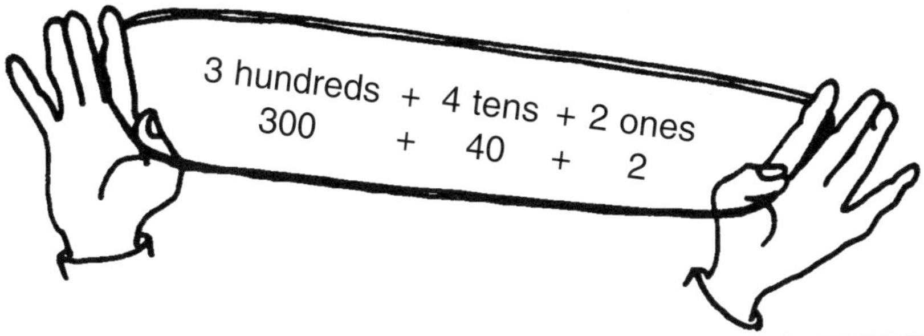

Build blocks. Draw pictures. Write each number in expanded notation.

1. 214

Hundreds	Tens	Ones
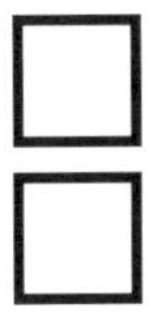		

_______ + _______ + _______

2. 176

Hundreds	Tens	Ones

_______ + _______ + _______

3. 131

Hundreds	Tens	Ones

_______ + _______ + _______

4. 243

Hundreds	Tens	Ones

_______ + _______ + _______

Write the number.

5. 900 + 50 + 1 = __________

6. 400 + 20 + 7 = __________

7. 600 + 10 + 9 = __________

8. 300 + 90 = __________

One Thousand

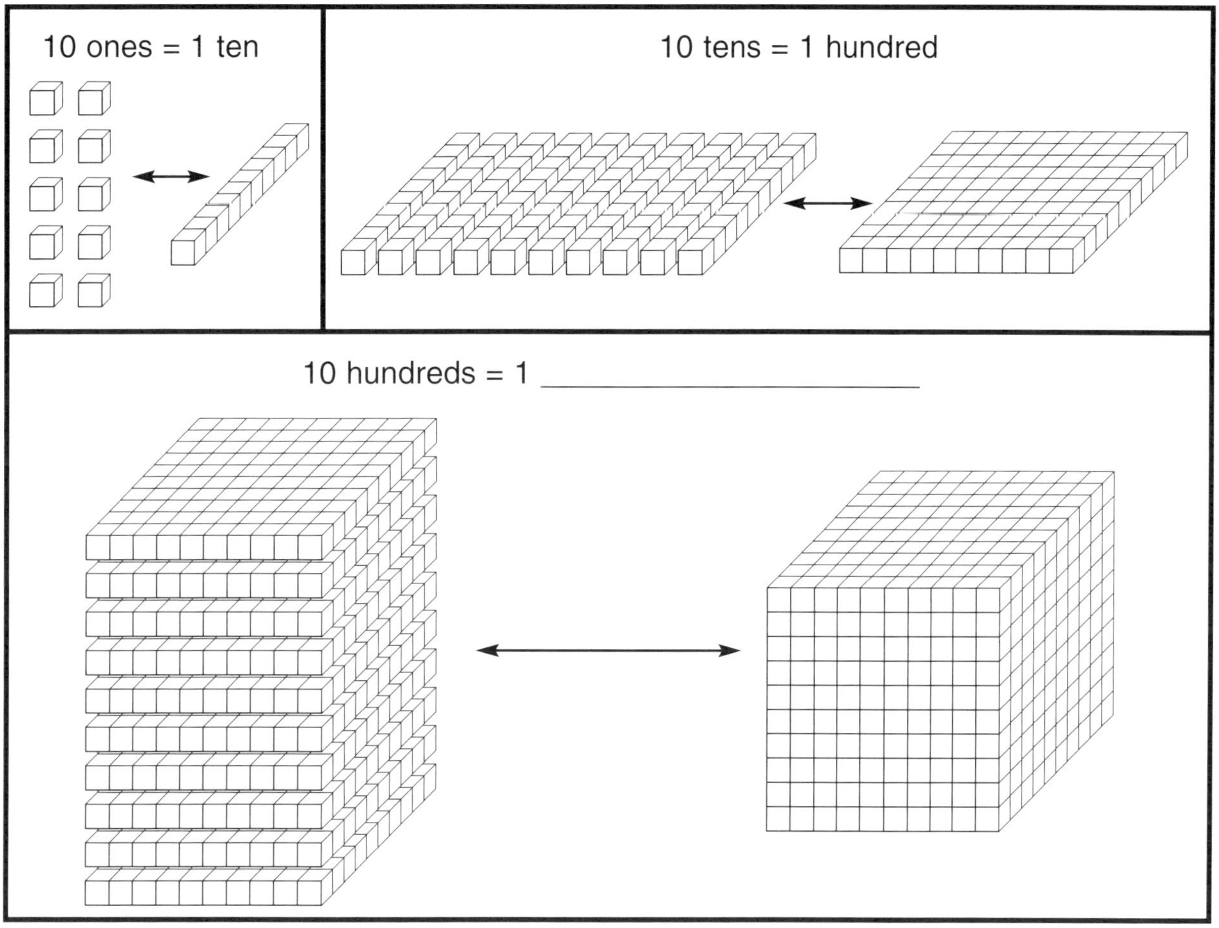

Fill in the blanks.

1. 10, 20, _____, _____, _____, _____, _____, _____, _____, 100

2. 310, 320, _____, _____, _____, _____, _____, _____, _____, 400

3. 100, 200, _____, _____, _____, _____, _____, _____, _____, 1000

4. 406, 407, _____, _____, _____, 411

5. 798, 799, _____, _____, _____, 803

Place Value in 6-Digit Numbers

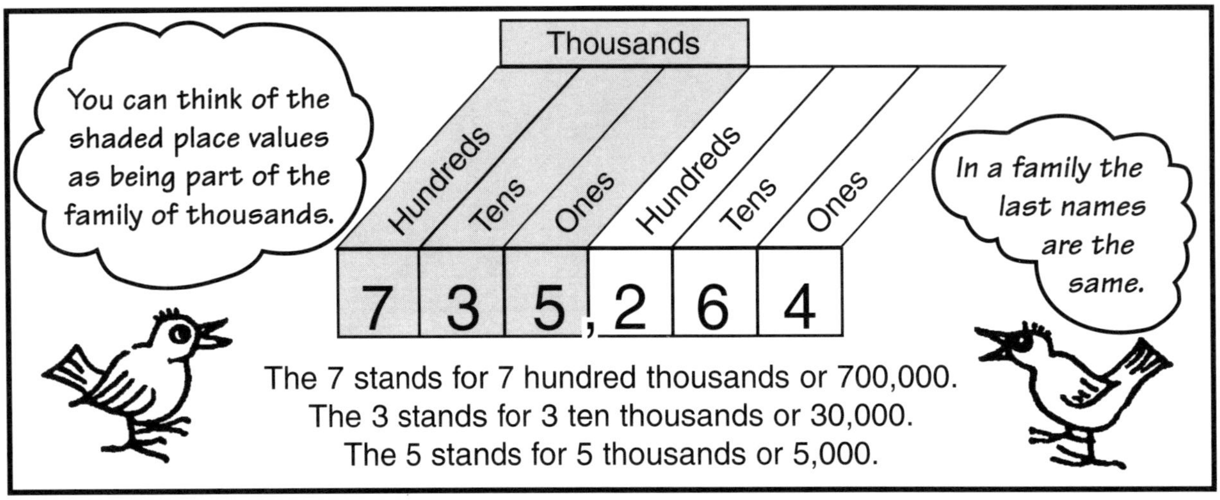

1. Write the name of the place value of each digit in the number 936,425.

4 ____________________ 6 ____________________ 5 ____________________

9 ____________________ 2 ____________________ 3 ____________________

Write the place name for each underlined digit.

2. $240,\underline{7}63$ ____________________ **3.** $23\underline{4},569$ ____________________

4. $3\underline{6}5,492$ ____________________ **5.** $609,45\underline{3}$ ____________________

6. $210,4\underline{5}7$ ____________________ **7.** $\underline{9}45,362$ ____________________

8. $30\underline{4},765$ ____________________ **9.** $5\underline{3}0,612$ ____________________

10. What digit is in the hundreds place in 4527? ______

11. What digit is in the ten thousands place in 356,243? ______

12. What digit is in the thousands place in 713,406? ______

13. What digit is in the hundred thousands place in 507,612? ______

14. What digit is in the hundreds place in 314, 527? ______

15. What digit is in the ten thousands place in 517,463? ______

16. What digit is in the hundred thousands place in 430,628? ______

Ordering Numbers

Use < or > to compare the numbers.

1. 25 ☐ 36 **2.** 47 ☐ 21 **3.** 432 ☐ 470

4. 532 ☐ 510 **5.** 400 ☐ 250 **6.** 379 ☐ 412

7. 698 ☐ 689 **8.** 535 ☐ 545 **9.** 1257 ☐ 2259

---

10. Which number is least? 3246 3400 3602 3200 ________

11. Which number is greatest? 6789 6712 6730 6790 ________

12. Which numbers are greater than 4632? 4215 5493 4623 4658.

13. Which numbers are greater than 4509? 4387 4590 4519 4500.

Order the numbers from least to greatest.

14. 330 307 323 315 328

_____ _____ _____ _____ _____

15. 783 688 694 649 790

_____ _____ _____ _____ _____

16. 5000 3000 4000 6000

_____ _____ _____ _____

17. 7394 6257 9000 5836

_____ _____ _____ _____

18. 3500 3700 3100 3800

_____ _____ _____ _____

19. 6314 6308 6379 6357

_____ _____ _____ _____

20. 5749 5342 5600 5794

_____ _____ _____ _____

21. 2148 2109 2384 2097

_____ _____ _____ _____

22. 3489 2176 3984 2764

_____ _____ _____ _____

23. 1023 950 1197 1950

_____ _____ _____ _____

MAKE THE LARGEST NUMBER

	Thousand	Hundred	Ten	One	Throw Away
1					
2					
3					

MAKE THE SMALLEST NUMBER

	Thousand	Hundred	Ten	One	Throw Away
1					
2					
3					

Reading and Writing 4-Digit Numbers

Write the numeral for each word name.

1. five ______

2. fifty ______

3. five hundred ______

4. five thousand ______

5. sixty-two ______

6. twelve ______

7. three hundred twenty-seven ______

8. nine hundred nineteen ______

9. seven hundred six ______

10. eight hundred eleven ______

11. three hundred forty ______

12. five hundred sixty ______

13. two thousand five hundred forty-seven ______

14. eight thousand six hundred twenty-three ______

15. seven thousand one hundred twenty-two ______

16. one thousand two hundred fifty ______

17. seven thousand one hundred two ______

18. four thousand seventy ______

19. six thousand thirty-five ______

20. eight thousand ten ______

21. Write the largest 4-digit number. ______________

22. Write the smallest 4-digit number. ______________

Counting by 3's, 4's and 6's

1. To count by 3's, start at 0. Count over 3 to 3. Count over 3 more to 6, etc.
What numbers do you land on when you count by 3's?
Draw a triangle around the landing points in the chart below.

0 _____ _____ _____ _____ _____

_____ _____ _____ _____ 30

2. To count by 4's, start at 0. Count over 4 to 4. Count over 4 more to 8, etc.
What numbers do you land on when you count by 4's?
Draw a circle around the landing points in the chart below.

0 _____ _____ _____ _____ _____

_____ _____ 32 _____ _____

1	2	3	4	5	6	7	8	9	10
11	12	13	14	15	16	17	18	19	20
21	22	23	24	25	26	27	28	29	30
31	32	33	34	35	36	37	38	39	40
41	42	43	44	45	46	47	48	49	50
51	52	53	54	55	56	57	58	59	60
61	62	63	64	65	66	67	68	69	70
71	72	73	74	75	76	77	78	79	80
81	82	83	84	85	86	87	88	89	90
91	92	93	94	95	96	97	98	99	100

3. Which numbers were landing points for the 3's and the 4's? (Hint: circles & triangles?)

____

4. Count by 6's.

0 ____ ____ ____

____ ____ ____ 42

____ ____ ____

Nearly 3,000 workers have been let go since the first of the year.

500,000 Fans Attend World Cup Match

40,000 CHILL OUT AT WINTER CARNIVAL!

All candy on sale for 27¢ apiece.

Rounding to the Nearest Ten

Many numbers seen in print are obtained by rounding. Rounding is helpful in making quick comparisons and approximate calculations.

Example: Round 27 to the nearest ten.

I built 27 with base ten blocks. Then I built 20 and 30.

I know 27 is closer to 30 than 20 because

____

____

____

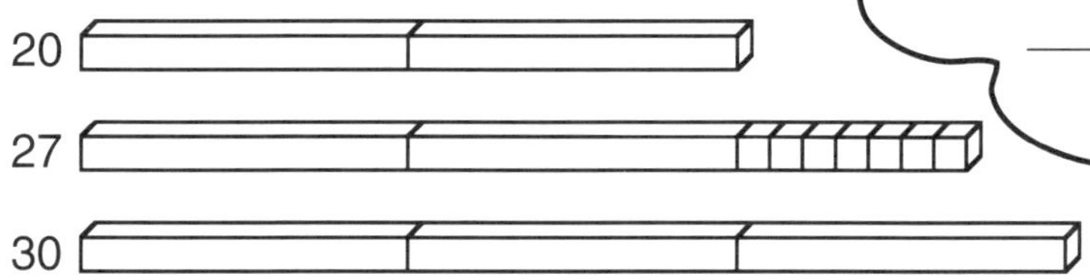

Use base ten blocks to round each number to the nearest ten. Look for a pattern.

1. 23 ______	**2.** 29 ______	**3.** 24 ______	**4.** 25 ______
5. 42 ______	**6.** 81 ______	**7.** 96 ______	**8.** 67 ______
9. 12 ______	**10.** 18 ______	**11.** 64 ______	**12.** 15 ______
13. 131 ______	**14.** 265 ______	**15.** 146 ______	**16.** 482 ______
17. 279 ______	**18.** 682 ______	**19.** 630 ______	**20.** 998 ______
21. 1321 ______	**22.** 1483 ______	**23.** 1157 ______	**24.** 1516 ______

25. Describe the pattern for rounding to the nearest ten.

____

____

____

____

Rounding to the Nearest Hundred

You can use base ten blocks and a "halfway" number to help round to the nearest 100.

Example: Round 482 to the nearest hundred.

1. Build the number 482.

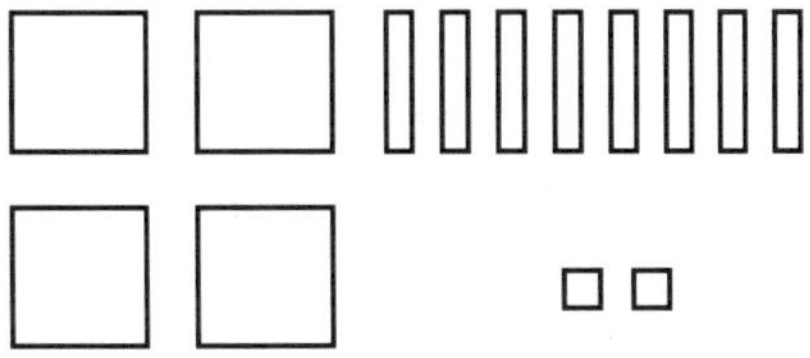

2. Build the two groups of hundreds between which 482 falls.

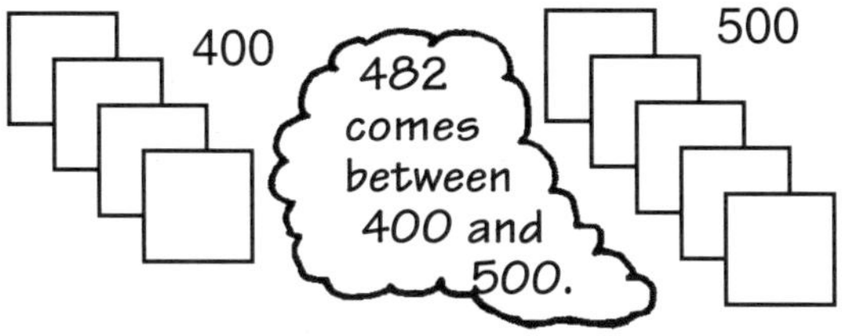

3. Compare 482 to 400 and to 500 by thinking of a "halfway" number.

450 is the halfway number between 400 and 500.
482 is more than the halfway number.
482 rounds to 500

Round each number to the nearest 100.

1. 326 comes between _____ and _____.
The halfway number is ______.
326 is (greater, less) than 350.
326 rounds to ______.

2. 368 comes between _____ and _____.
The halfway number is ______.
368 is (greater, less) than 350.
368 rounds to ______.

	Number	Comes Between	Halfway Number	Rounds To
3.	618	_____ and _____	______	618 rounds to _____
4.	227			
5.	461			
6.	653			
7.	90			
8.	350			
9.	1517			
10.	1728			

The Order Property in Addition

A property is a rule that is always true. The order property for addition is illustrated.

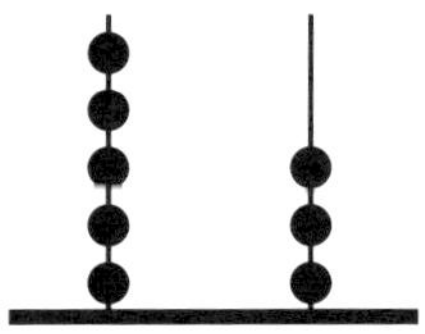

5 + 3 = 8

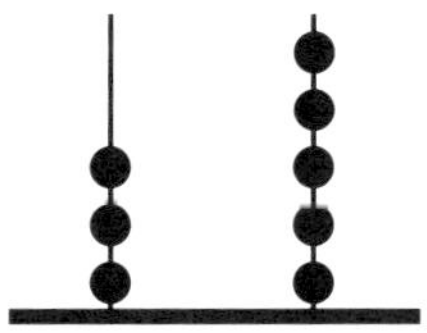

3 + 5 = 8

Show the addition problems on the number line and write the answer.

1. 4 + 3 =

2. 3 + 4 =

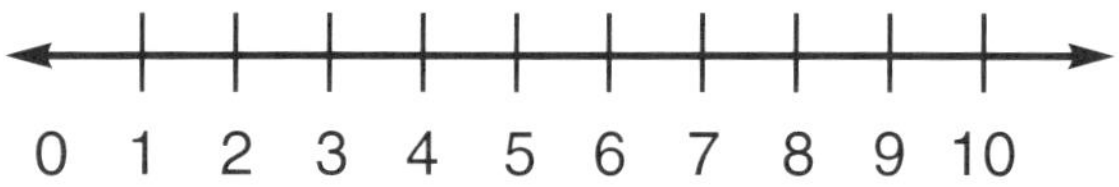

3. How do the answers in problems 1 and 2 above compare?

4. Since 8 + 7 = 15,
7 + 8 = _____

5. Since 6 + 5 = 11,
5 + 6 = _____

6. Since 9 + 3 = 12,
3 + 9 = _____

7. Since 7 + 8 = 15,
7 + 8 = _____

8. Since 43 + 16 = 59,
16 + 43 = _____

9. Since 38 + 75 = 113,
75 + 38 = _____

What number goes in the box?

10. 7 + 4 = ☐ + 7

11. 8 + 3 = ☐ + 8

12. 6 + ☐ = 5 + 6

13. ☐ + 7 = 7 + 8

14. 5 + 20 = ☐ + 5

15. 13 + 7 = 7 + ☐

16. 23 + 46 = 46 + ☐

17. 42 + 35 = ☐ + 42

18. Describe in your own words the order property of addition.

19. Give an example of the order property of addition.

The Grouping Property of Addition

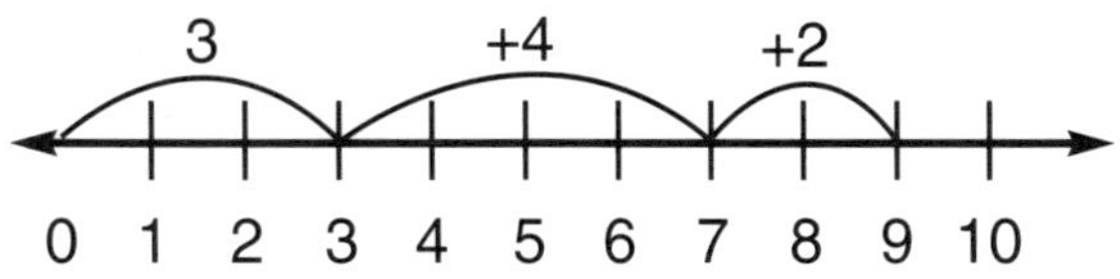

(3 + 4) + 2 =

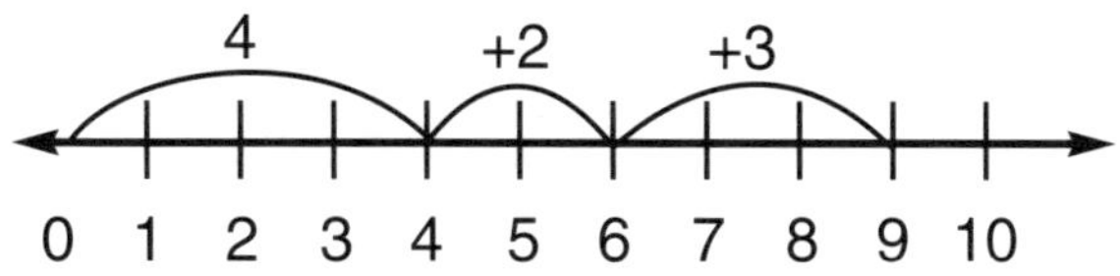

(4 + 2) + 3 =

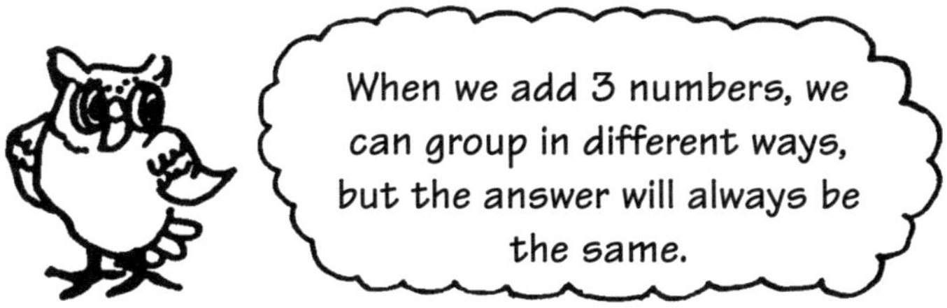

Show the addition problems on the number lines and write the answers.

1. (2 + 5) + 1 =

2. (2 + 1) + 5 =

3. How do the answers in problems 1 and 2 above compare?

4. (6 + 5) + 4 = ______

(6 + 4) + 5 = ______

5. (23 + 12) + 7 = ______

(23 + 7) + 12 = ______

6. 12 + (45 + 28) = ______

(12 + 28) + 45 = ______

7. 93 + (125 + 7) = ______

(93 + 7) + 125 = ______

8. (2 + 4) + 3 = 2 + (4 + $\square$)

9. (9 + 6) + 4 = 9 + ($\square$ + 4)

10. (13 + 7) + 5 = 13 + (7 + $\square$)

11. (15 + 7) + 3 = $\square$ + (7 + 3)

12. 20 + (7 + 30) = (20 + $\square$) + 30

13. 14 + (12 + 8) = (14 + 12) + $\square$

14. (90 + 38) + 2 = 90 + (38 + $\square$)

15. (50 + 20) + 3 = $\square$ + (20 + 3)

16. Explain in your own words the grouping property of addition. ______________

____

17. Give an example of the grouping property of addition. ______________

____

Adding 3-Digit Numbers, Regrouping from Ones to Tens

Find the sums.

1.	235 + 146 = ☐	**2.**	329 + 252 = ☐	**3.**	404 + 266 = ☐	**4.**	568 + 123 = ☐
5.	256 + 636	**6.**	487 + 103	**7.**	408 + 351	**8.**	178 + 512
9.	397 + 487	**10.**	285 + 189	**11.**	473 + 282	**12.**	598 + 399

Sneaky Pete caught these mistakes. What error did each student make?

A 538 + 217 = 791	**B** 538 + 217 = 845	**C** 538 + 217 = 321	**D** 538 + 217 = 745

Adding Columns of Numbers

```
  24 }
  31 }  4 + 1 = 5
+ 16      ↓
  71
```

```
Check
  71
  ==
  24      ↑
  31 }
  16 }  6 + 1 = 7
```

Find the sums.

1.

T	O
2	4
3	0
+ 4	1

2.

T	O
2	3
1	4
+ 5	2

3.

T	O
4	2
1	1
2	2
+ 1	3

4.

T	O
2	3
1	0
3	1
+ 2	4

5. $\begin{array}{r} 25 \\ 13 \\ +\ 34 \\ \hline \end{array}$

6. $\begin{array}{r} 40 \\ 27 \\ +\ 16 \\ \hline \end{array}$

7. $\begin{array}{r} 23 \\ 12 \\ 41 \\ +\ 15 \\ \hline \end{array}$

8. $\begin{array}{r} 16 \\ 50 \\ 13 \\ +\ 24 \\ \hline \end{array}$

Add. Check by adding up.

9. $\begin{array}{r} 25 \\ 13 \\ +\ 41 \\ \hline \end{array}$

10. $\begin{array}{r} 34 \\ 21 \\ +\ 37 \\ \hline \end{array}$

11. $\begin{array}{r} 50 \\ 26 \\ +\ 19 \\ \hline \end{array}$

12. $\begin{array}{r} 27 \\ 48 \\ +\ 13 \\ \hline \end{array}$

Adding 4-Digit Numbers

	Th	H	T	O
	1	3	4	5
+	1	8	0	3
	2	11	4	8
	3	1	4	8

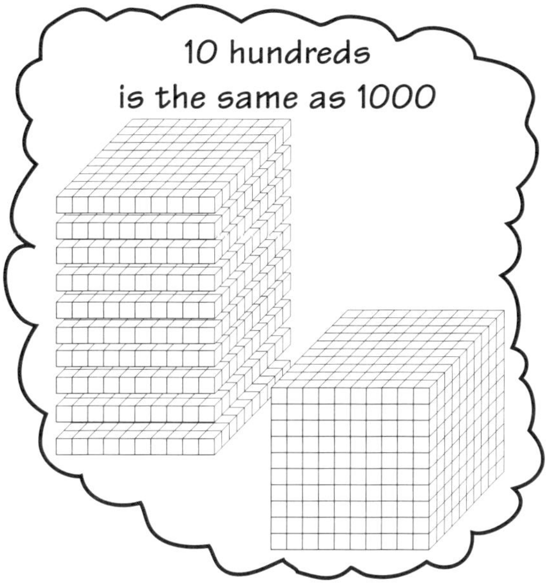

Shortcut:

$$\begin{array}{r} {\scriptstyle 1} \\ 1345 \\ +\ 1803 \\ \hline 3148 \end{array}$$

Find the sums.

1.

	Th	H	T	O
	1	8	6	5
+	1	4	3	1

Start here.

2.

	Th	H	T	O
	1	6	1	2
+	1	9	4	3

3.

	Th	H	T	O
	1	5	8	3
+	1	5	0	4

4. $\begin{array}{r} 5731 \\ +\ 3914 \\ \hline \end{array}$

5. $\begin{array}{r} 3750 \\ +\ 2746 \\ \hline \end{array}$

6. $\begin{array}{r} 2935 \\ +\ 4923 \\ \hline \end{array}$

Watch out! More than one exchange.

7. $\begin{array}{r} 2783 \\ +\ 4608 \\ \hline \end{array}$

8. $\begin{array}{r} 2528 \\ +\ 3937 \\ \hline \end{array}$

9. $\begin{array}{r} 4683 \\ +\ 3691 \\ \hline \end{array}$

Horizontal Addition

You can use graph paper to help line up numbers correctly.

744 + 36 + 1431 = ?

Th	H	T	O
	7	4	4
		3	6
1	4	3	1

Line up numbers under the ones on the right side.

Start adding here.

Rewrite each problem on graph paper. Add.

1.
123 + 87 + 1361 =

Th	H	T	O

2.
294 + 3026 + 17 =

Th	H	T	O

3.
2346 + 24 + 603 =

Th	H	T	O

4.
924 + 29+ 8 =

Th	H	T	O

5.
753 + 3146 + 48 =

Th	H	T	O

6.
23,410 + 56 + 3,595 =

TTh	Th	H	T	O

Sneaky Pete found one student who got 11,291 for problem 1. What error was made?

Subtracting 3-Digit Numbers

542
− 325

First, build the larger number.
Then remove the smaller number.

542

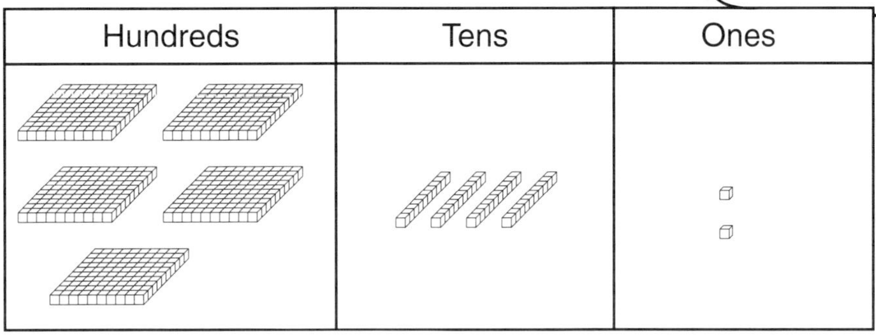

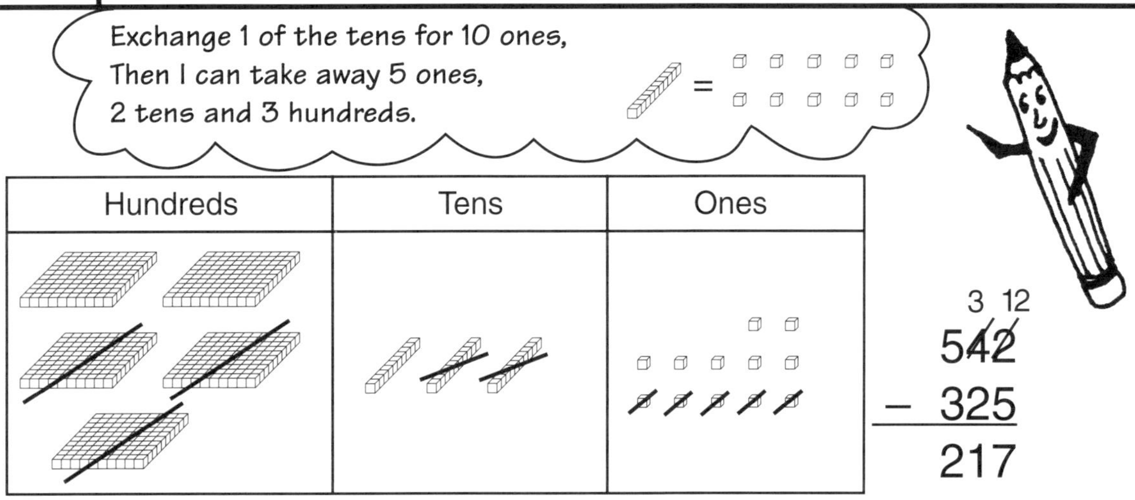

$$\begin{array}{r} \overset{3}{5}\,\overset{12}{\not4\not2} \\ -\ 325 \\ \hline 217 \end{array}$$

Subtract. Remember to show more ones.

1.	$\begin{array}{r} 342 \\ -\ 216 \\ \hline \end{array}$	**2.**	$\begin{array}{r} 571 \\ -\ 238 \\ \hline \end{array}$	**3.**	$\begin{array}{r} 730 \\ -\ 417 \\ \hline \end{array}$	**4.**	$\begin{array}{r} 678 \\ -\ 339 \\ \hline \end{array}$
5.	$\begin{array}{r} 493 \\ -\ 345 \\ \hline \end{array}$	**6.**	$\begin{array}{r} 929 \\ -\ 341 \\ \hline \end{array}$	**7.**	$\begin{array}{r} 739 \\ -\ 366 \\ \hline \end{array}$	**8.**	$\begin{array}{r} 314 \\ -\ 206 \\ \hline \end{array}$

Sneaky Pete found these answers for problems 1–4.
What error was he making?
***1.** 134* ***2.** 347* ***3.** 327* ***4.** 341*

Subtracting 3-Digit Numbers, Two Regroupings

Subtract. You will have to regroup twice.

1. 431 − 158 = ____

2. 746 − 57 = ____

3. 428 − 239 = ____

4. 751 − 263 = ____

5. 834 − 656 = ____

6. 932 − 344 = ____

7. 740 − 367 = ____

8. 791 − 584 = ____

Sneaky Pete found these answers for problems 1–4.
What error was the student making?

1. 373 **2.** 789 **3.** 289 **4.** 588

Subtracting Across Zeros

Subtract.

1.	202 – 85	**2.**	305 – 148	**3.**	501 – 362	**4.**	703 – 236
5.	608 – 239	**6.**	804 – 257	**7.**	400 – 23	**8.**	300 – 176
9.	500 – 182	**10.**	504 – 46	**11.**	504 – 249	**12.**	605 – 50

Subtracting 4-Digit Numbers

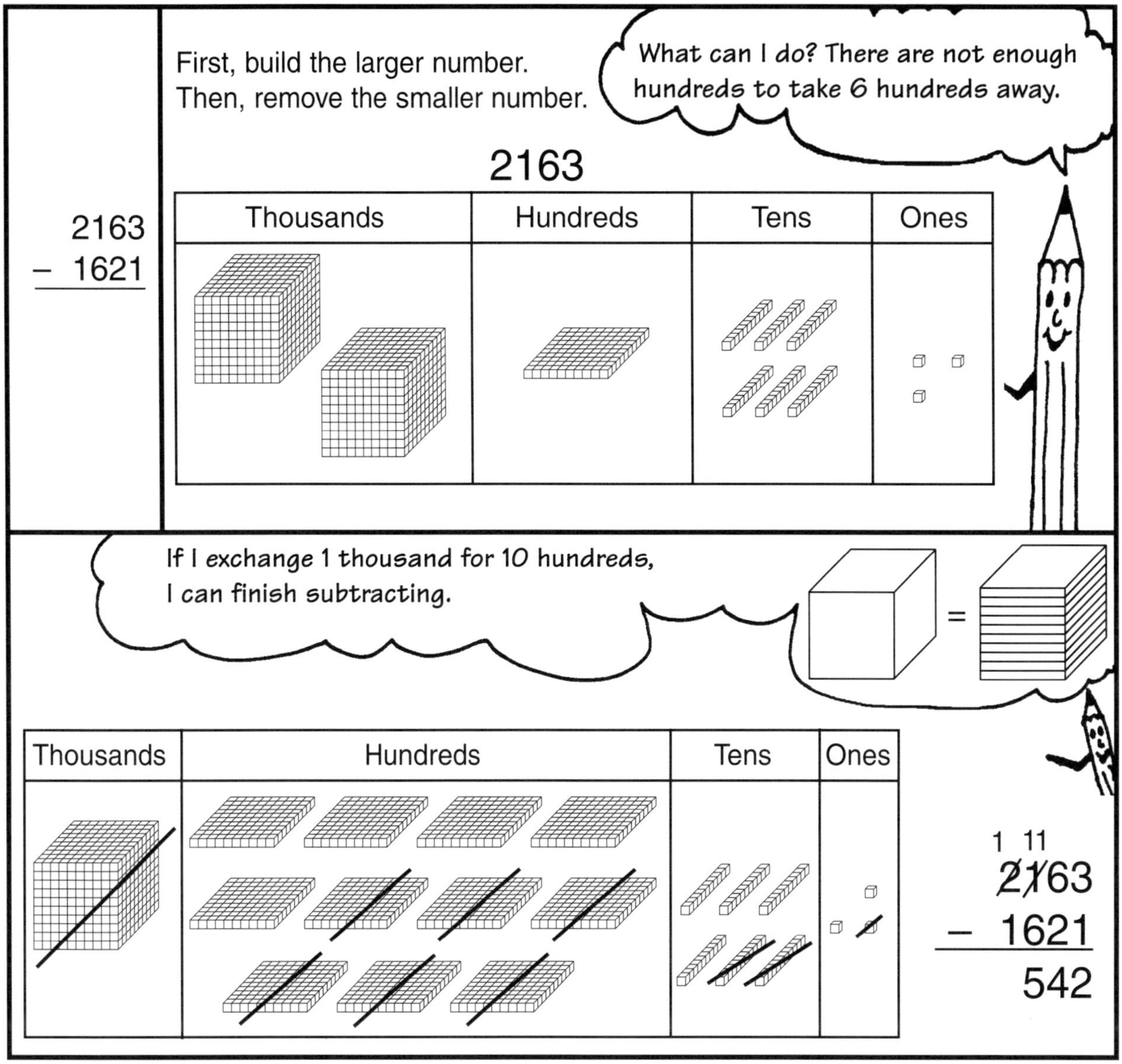

Subtract.

1.	2473 − 1652	**2.**	2369 − 1716	**3.**	2578 − 1934	**4.**	3287 − 1635
5.	6171 − 3725	**6.**	7292 − 4807	**7.**	6938 − 2492	**8.**	7876 − 1283

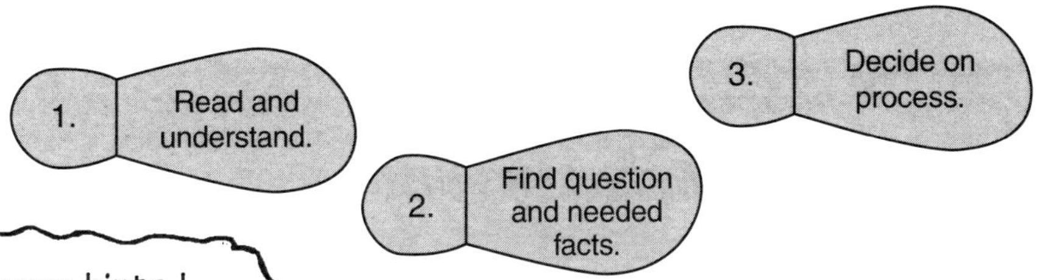

Decide on a Process

If you:
put things together—add
take things away—subtract
compare things—subtract

Decide if you should add or subtract to solve these problems.
Circle the correct answer.

1. Sara has 17 play cars and Jane has 10 play cars. How many cars do they have in all?

add subtract

2. Tom baked 48 cookies. He gave 25 away. How many cookies does Tom have left?

add subtract

3. A lion weighed 826 pounds and a gorilla weighed 670 pounds. How much more did the lion weigh?

add subtract

4. Kia earned $35. She already had $54. How much money does she have altogether?

add subtract

5. Pedro is 45 inches tall and Marcy is 47 inches tall. What is their total height?

add subtract

6. Linda got 87 correct on her test. Magda got 62 correct on her test. How many more did Linda get correct?

add subtract

7. Leslie sold 115 tickets and Leroy sold 124 tickets. How many tickets did they sell in all?

add subtract

8. James weighed 130 pounds. He lost 12 pounds. How much did he weigh after his diet?

add subtract

Estimating Sums

4. Estimate.

Approximate numbers are often used to estimate an answer. Rounded numbers are approximate numbers.

$$\begin{array}{r} 623 \\ +\ 792 \\ \hline \end{array}$$

Round each number to the leading digit.

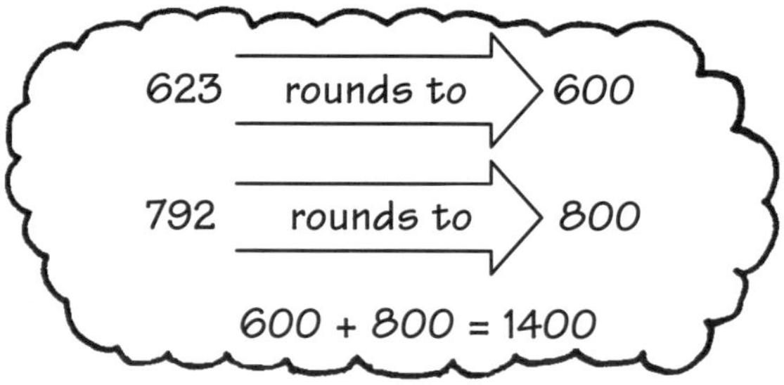

1. Which will give the best estimate for 81 + 68?

A 80 + 60 **B** 90 + 70 **C** 70 + 70 **D** 80 + 70

2. Which will give the best estimate for 427 + 684?

A 400 + 600 **B** 500 + 700 **C** 300 + 700 **D** 400 + 700

Use estimated sums to complete each table.

3.

+	21	59
38		
52		
47		

4.

+	32	66
17		
85		
74		

5.

+	370	637
210		
380		
450		

6.

+	420	590
364		
274		
158		

7.

+	1815	6184
4216		
3724		
2549		

8.

+	1827	4281
5178		
3547		
7862		

Add or Subtract?

These problems do not have numbers in them. Find the clue words to help decide if you should add or subtract.

1. Mrs. Fuller sent _____ Christmas cards this year. She received _____. How many more did she send?

subtract

2. There were _____ sunny days in June and _____ in July. How many days were sunny in all?

3. The bakery had _____ loaves of bread in the morning. At the end of the day, there were still _____ loaves of bread. How many loaves were sold?

4. This year _____ boys played baseball. Last year _____ boys played. How many more played this year?

5. Kara has _____ books on her shelf. At the library she gets _____ more. How many does she have altogether?

6. Bob scored _____ points in the basketball game this week. Last week he scored _____. How many more did he score this week?

7. Candy had _____ outfits for her Barbie doll. She bought _____ more. How many does she have in all?

8. Ted had _____ candy bars to sell for the school fair. Today he sold _____. How many are left to sell?

9. Mrs. Johnson bought _____ cases of pop for the dance. There were _____ left. How many cases were sold?

10. _____ boys and _____ girls were at the party. How many were there in all?

Adding and Subtracting Money

100¢ + 50¢ + 30¢ = 180¢ or $1.80

Write the value of the money, using the ¢ sign and the $ sign.

1. $1 bill, 1 quarter, 2 dimes = ________¢ or $__________
2. $1 bill, 4 dimes, 1 nickel, 4 pennies = ________¢ or $__________
3. $1 bill, 1 half-dollar, 1 dime, 7 pennies = ________¢ or $__________
4. $5 bill, $1 bill, 3 quarters, 1 dime = ________¢ or $__________

Add the amounts.

5.	$ 2.45 + 3.10	**6.**	$ 4.75 + 3.25	**7.**	$ 2.65 + 0.86	**8.**	$ 12.45 + 7.25
9.	$ 0.68 + 0.58	**10.**	$ 1.56 + 0.49	**11.**	$ 0.48 + 0.56	**12.**	$ 1.75 + 2.48

Subtract the amounts.

13.	$ 3.75 − 1.42	**14.**	$ 4.60 − 2.00	**15.**	$ 6.75 − 0.15	**16.**	$ 3.65 − 1.37
17.	$ 1.00 − 0.35	**18.**	$ 5.00 − 0.25	**19.**	$ 5.00 − 1.48	**20.**	$ 10.00 − 4.27

What coins are in each bank?

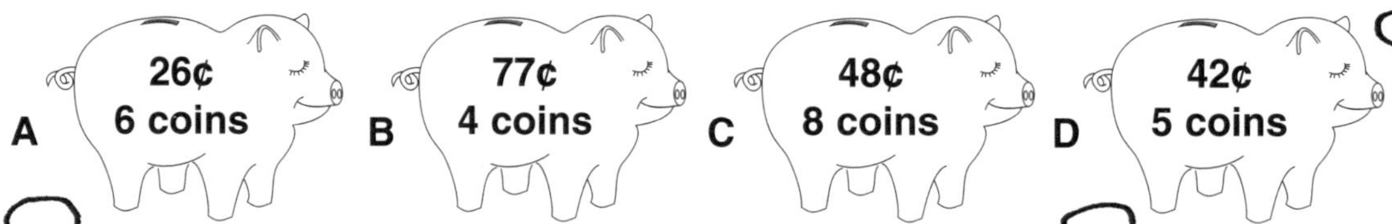

Multiplication Facts

Multiplication facts can be found by skip counting on a number line.

2 groups of 4 become 2 jumps of 4 on a number line.
2 groups of 4 = 8. 2 jumps of 4 lands on 8.

Use skip counting to solve the following problems.

2's

2	2	2	2	2	2	2	2	2
× 1	× 2	× 3	× 4	× 5	× 6	× 7	× 8	× 9

3's

3	3	3	3	3	3	3	3	3
× 1	× 2	× 3	× 4	× 5	× 6	× 7	× 8	× 9

4's

4	4	4	4	4	4	4	4	4
× 1	× 2	× 3	× 4	× 5	× 6	× 7	× 8	× 9

5's

5	5	5	5	5	5	5	5	5
× 1	× 2	× 3	× 4	× 5	× 6	× 7	× 8	× 9

Mixed

5	3	4	2	5	3	4	3	4
× 9	× 8	× 9	× 9	× 7	× 9	× 8	× 7	× 7

Building Multiplication Facts

You can use blocks to build a square or rectangle of each of the multiplication facts.

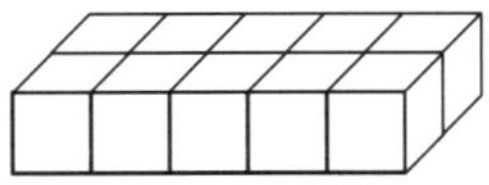

This rectangle has 5 units on one side and 2 units on the other side.

2 groups of 5 = 10

$2 \times 5 = 10$

You can outline your rectangle on graph paper to draw a picture of a multiplication fact.

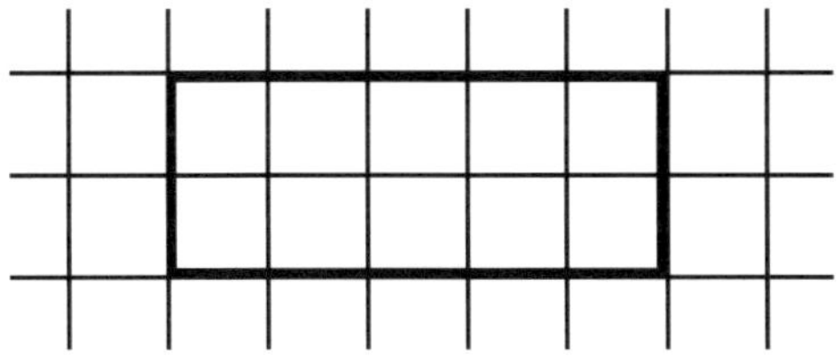

2 rows of 5 = 10

$2 \times 5 = 10$

Use blocks to build a model of each multiplication fact.
Draw a picture of each fact on graph paper.

1. $3 \times 5 =$ **2.** $4 \times 5 =$ **3.** $6 \times 5 =$ **4.** $7 \times 5 =$

5. $2 \times 4 =$ **6.** $2 \times 7 =$ **7.** $2 \times 9 =$ **8.** $2 \times 6 =$

What multiplication fact is shown?

9. 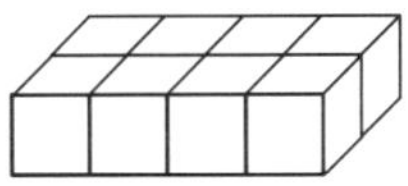

10. 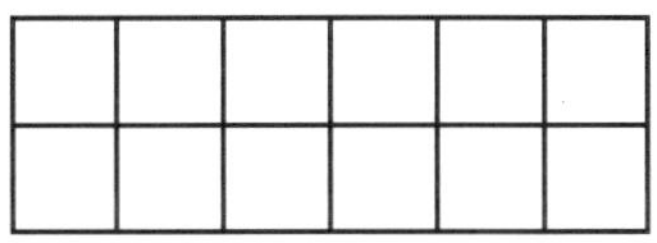

11. Use base ten blocks and graph paper to build a model and draw a picture of the multiplication facts with 3's.

12. Use base ten blocks and graph paper to build a model and draw a picture of the multiplication facts with 4's.

Mental Math

To find products of 9, think of multiplying by 10.

9 × 6 = 10 sixes less 1 six
= 60 – 6 or 54

9 × 7 = 10 sevens less 1 seven
= 70 – 7 or 63

1.

×9	0	1	2	3	4	5	6	7	8	9	10

Think of squares of numbers as special products.

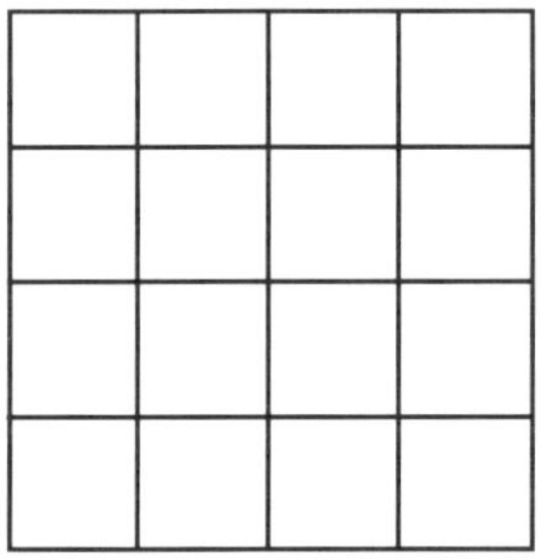

4 × 4 = 16

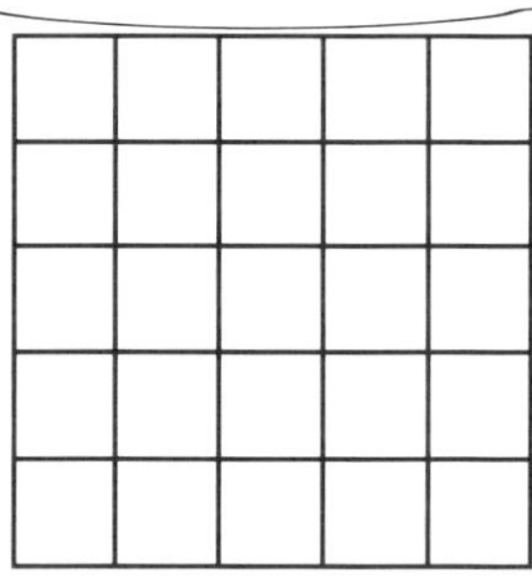

5 × 5 = 25

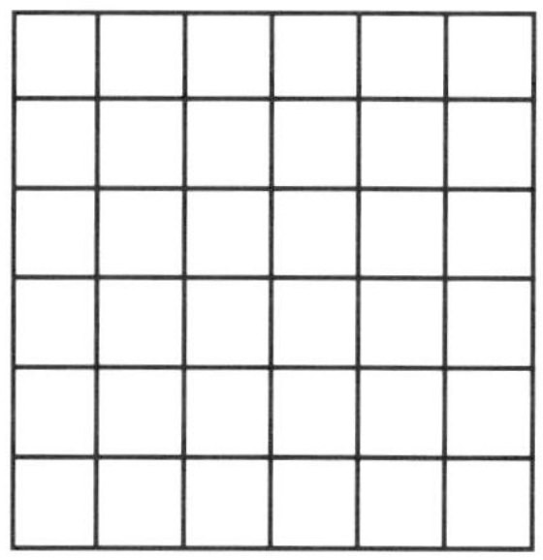

6 × 6 = 36

2.

1	2	3	4	5	6	7	8	9
× 1	× 2	× 3	× 4	× 5	× 6	× 7	× 8	× 9

Use the product of squares to discover other facts.
6 × 7 = the square of 6 plus 6 or 36 + 6 or 42.
6 × 8 = the square of 7 less 1 or 49 – 1 = 48.

3.

6	5	8	9	8	3	7	6	7
× 4	× 9	× 9	× 7	× 7	× 5	× 7	× 6	× 5

8	5	3	7	6	4	4	7	7
× 6	× 6	× 3	× 6	× 8	× 5	× 6	× 8	× 9

Multiplying 10's and Groups of 10's

Louie earns 20¢ a week for sorting cans and bottles for recycling.
How much does Louie earn in 3 weeks?

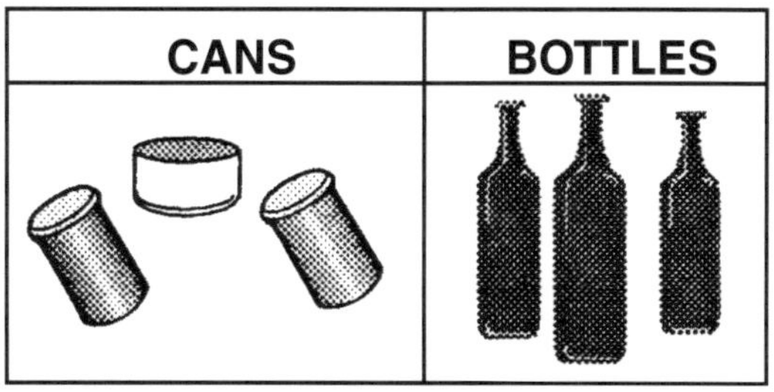

$$\begin{array}{r} 20¢ \\ 20¢ \\ +\ 20¢ \\ \hline 60¢ \end{array}$$

You can use base ten blocks to multiply and record the result.

$$\begin{array}{r} 20 \\ \times\ 3 \\ \hline 60 \end{array}$$

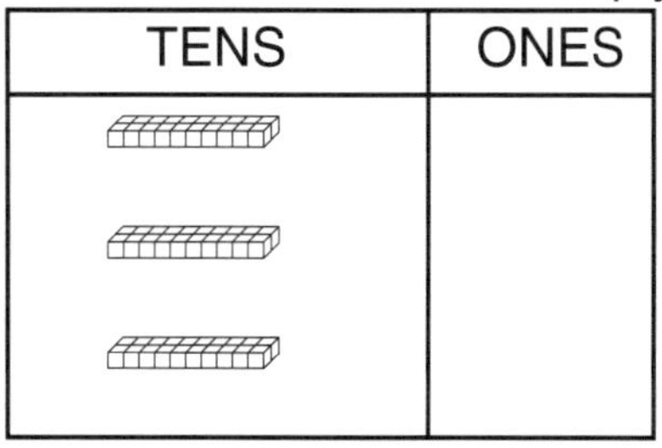

3 sets of 20 is 6 tens and 0 ones or 60.

$$\begin{array}{r} 20 \\ \times\ 3 \\ \hline 60 \end{array}$$

Use coins or blocks to multiply.

1. $\begin{array}{r} 20¢ \\ \times\ 4 \\ \hline \end{array}$ **2.** $\begin{array}{r} 10¢ \\ \times\ 6 \\ \hline \end{array}$ **3.** $\begin{array}{r} 10¢ \\ \times\ 8 \\ \hline \end{array}$ **4.** $\begin{array}{r} 30¢ \\ \times\ 3 \\ \hline \end{array}$

5. $\begin{array}{r} 20¢ \\ \times\ 5 \\ \hline \end{array}$ **6.** $\begin{array}{r} 30¢ \\ \times\ 4 \\ \hline \end{array}$ **7.** $\begin{array}{r} 40¢ \\ \times\ 4 \\ \hline \end{array}$ **8.** $\begin{array}{r} 50¢ \\ \times\ 3 \\ \hline \end{array}$

9. $\begin{array}{r} 60 \\ \times\ 4 \\ \hline \end{array}$ **10.** $\begin{array}{r} 90 \\ \times\ 3 \\ \hline \end{array}$ **11.** $\begin{array}{r} 70 \\ \times\ 5 \\ \hline \end{array}$ **12.** $\begin{array}{r} 80 \\ \times\ 4 \\ \hline \end{array}$

13. You earn 70¢ a week for folding laundry. How much do you earn in 4 weeks? ______

14. You save 50¢ one week and 40¢ the next week. How much money do you save? ______

Multiplying by 1 Digit, No Regrouping

Ong bought 3 dozen eggs at the market. How many eggs did she buy?

This problem has a hidden fact:
1 dozen = 12 eggs.

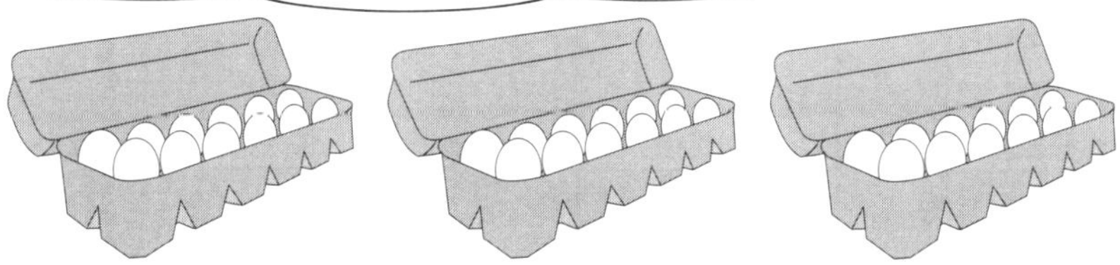

$$\begin{array}{r} 12 \\ 12 \\ +\ 12 \\ \hline 36 \end{array} \text{ eggs}$$

Multiplication is a shortcut for addition.

We can multiply to find the answer.

$$\begin{array}{r} 12 \\ \times\ 3 \\ \hline 36 \end{array} \text{ eggs}$$

1. $\begin{array}{r} 12 \\ \times\ 2 \\ \hline \end{array}$ **2.** $\begin{array}{r} 12 \\ \times\ 4 \\ \hline \end{array}$ **3.** $\begin{array}{r} 12 \\ \times\ 1 \\ \hline \end{array}$ **4.** $\begin{array}{r} 12 \\ \times\ 0 \\ \hline \end{array}$

5. $\begin{array}{r} 11 \\ \times\ 3 \\ \hline \end{array}$ **6.** $\begin{array}{r} 11 \\ \times\ 6 \\ \hline \end{array}$ **7.** $\begin{array}{r} 11 \\ \times\ 8 \\ \hline \end{array}$ **8.** $\begin{array}{r} 11 \\ \times\ 2 \\ \hline \end{array}$

9. $\begin{array}{r} 10 \\ \times\ 4 \\ \hline \end{array}$ **10.** $\begin{array}{r} 10 \\ \times\ 5 \\ \hline \end{array}$ **11.** $\begin{array}{r} 10 \\ \times\ 7 \\ \hline \end{array}$ **12.** $\begin{array}{r} 10 \\ \times\ 3 \\ \hline \end{array}$

13. Nick earned 1 dime every day for 11 weeks. How many dimes did he earn?

______

14. Ong had 3 dozen eggs. She used 1 dozen to bake an angel food cake. How many eggs did she have left?

______

Multiplying by 1 Digit with Regrouping

Use base ten blocks to find the multiplication pattern with regrouping.

Build 7 sets of 12.
Put together blocks having the same place value.

$$\begin{array}{r} 12¢ \\ \times\ 7 \\ \hline \end{array}$$

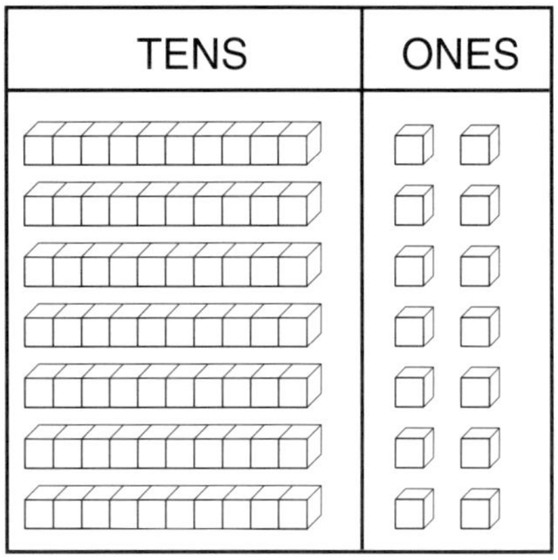

How many ones? 14 ones
Can you trade? Yes, trade 14 ones for 1 ten and 4 ones.

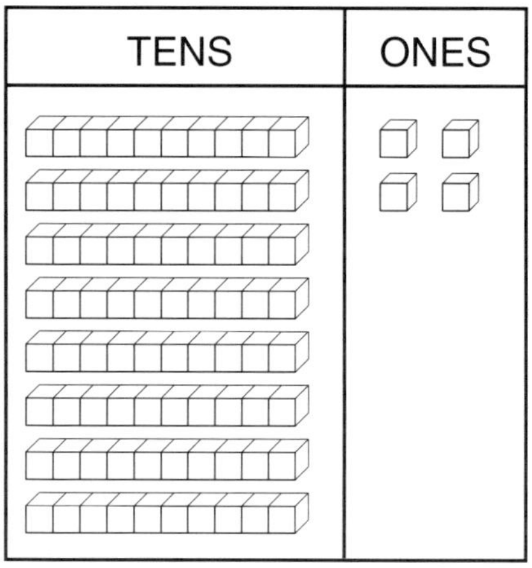

8 tens 4 ones = 84

The pattern: Multiply 7 times each place value block.
If possible, trade 10 of the same place value block for 1 of the next larger block.
Record:

$$\begin{array}{r} {}^{1} \\ 12¢ \\ \times\ 7 \\ \hline 4¢ \end{array}$$

7 times 2 ones is 14 ones.
14 ones is regrouped as 1 ten and 4 ones.

$$\begin{array}{r} {}^{1} \\ 12¢ \\ \times\ 7 \\ \hline 84¢ \end{array}$$

7 times 1 ten is 7 tens plus the 1 regrouped ten is 8 tens.

Multiply.

1. □ $\begin{array}{r} 15 \\ \times\ 5 \\ \hline \end{array}$
2. □ $\begin{array}{r} 24 \\ \times\ 3 \\ \hline \end{array}$
3. □ $\begin{array}{r} 36 \\ \times\ 2 \\ \hline \end{array}$
4. □ $\begin{array}{r} 14 \\ \times\ 6 \\ \hline \end{array}$
5. □ $\begin{array}{r} 17 \\ \times\ 4 \\ \hline \end{array}$
6. □ $\begin{array}{r} 26 \\ \times\ 3 \\ \hline \end{array}$
7. □ $\begin{array}{r} 34 \\ \times\ 2 \\ \hline \end{array}$
8. □ $\begin{array}{r} 35 \\ \times\ 2 \\ \hline \end{array}$
9. □ $\begin{array}{r} 23 \\ \times\ 3 \\ \hline \end{array}$
10. □ $\begin{array}{r} 16 \\ \times\ 5 \\ \hline \end{array}$
11. □ $\begin{array}{r} 24 \\ \times\ 4 \\ \hline \end{array}$
12. □ $\begin{array}{r} 49 \\ \times\ 2 \\ \hline \end{array}$

Multiplying a 3-Digit Number by a 1-Digit Number

Build 2 sets of 632. Put together blocks having the same place value.
There will be 4 ones, 6 tens and 12 hundreds.

$$\begin{array}{rl} 632 & \\ \times\ \ 2 & \\ \hline 4 & (2 \times 2) \\ 60 & (2 \times 30) \\ 1200 & (2 \times 600) \\ \hline 1264 & \end{array}$$

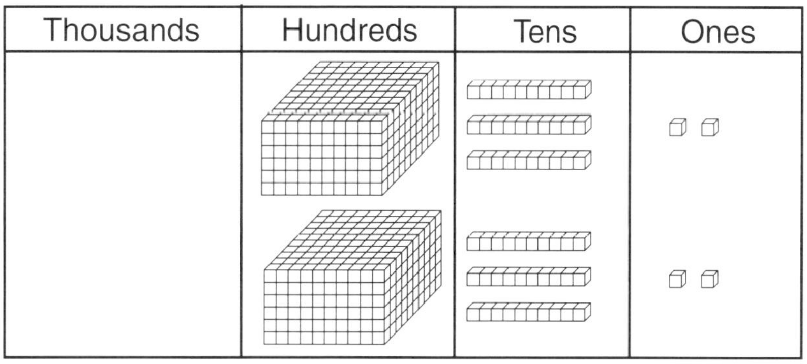

Trade 10 hundreds for 1 thousand.
There will be 4 ones, 6 tens, 2 hundreds
and 1 thousand...1264.

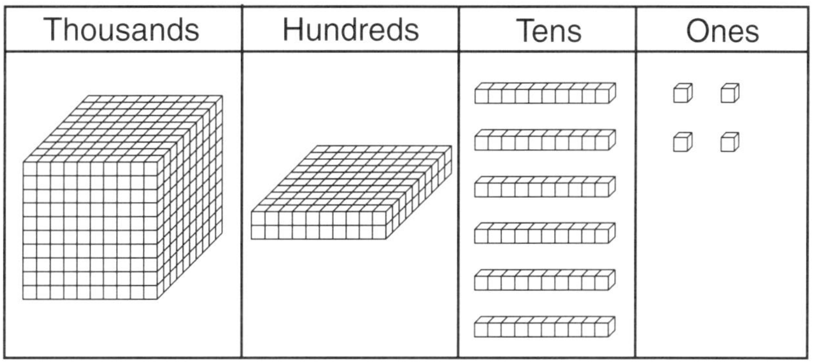

Record:

$$\begin{array}{r} 632 \\ \times\ \ 2 \\ \hline 1264 \end{array}$$

Multiply.

1. $\begin{array}{r} 312 \\ \times\ \ 4 \\ \hline \end{array}$ **2.** $\begin{array}{r} 523 \\ \times\ \ 3 \\ \hline \end{array}$ **3.** $\begin{array}{r} 814 \\ \times\ \ 2 \\ \hline \end{array}$ **4.** $\begin{array}{r} 313 \\ \times\ \ 3 \\ \hline \end{array}$

5. $\begin{array}{r} 623 \\ \times\ \ 4 \\ \hline \end{array}$ **6.** $\begin{array}{r} 842 \\ \times\ \ 3 \\ \hline \end{array}$ **7.** $\begin{array}{r} 146 \\ \times\ \ 5 \\ \hline \end{array}$ **8.** $\begin{array}{r} 457 \\ \times\ \ 4 \\ \hline \end{array}$

Using Base Ten Blocks to Multiply

Example: 2 × 162 = 2 × (100 + 60 + 2)

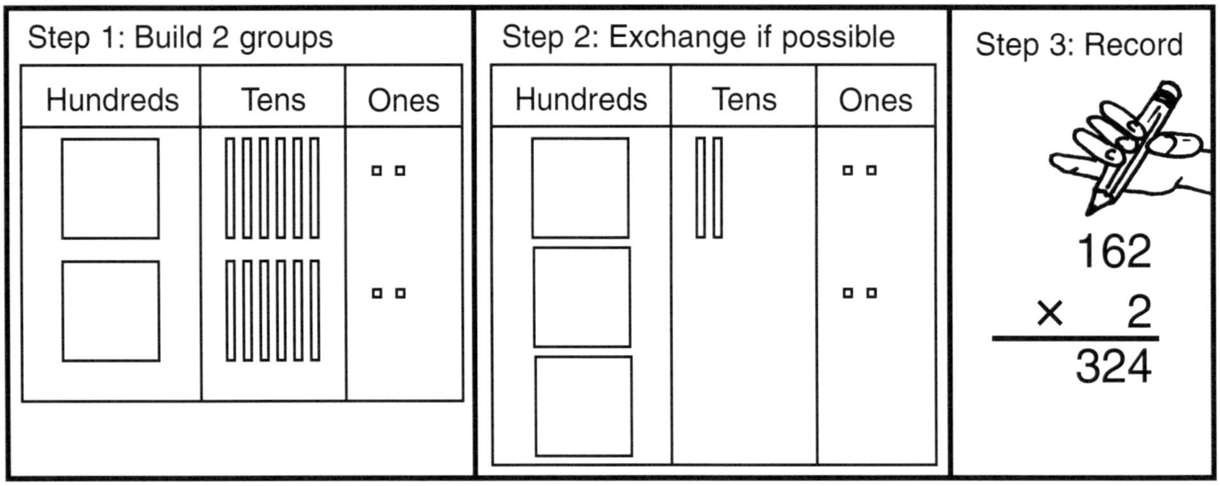

Use base ten blocks to multiply.

Draw a picture. (Use ▫ for ones, ▯ for tens and □ for hundreds.)

1. 4 × 132

Hundreds	Tens	Ones

Hundreds	Tens	Ones

Work Space

2. 3 × 153

Hundreds	Tens	Ones

Hundreds	Tens	Ones

Work Space

3. Make up a word problem about 4 × 132.

____

____

4. Make up a word problem about 3 × 153.

____

____

Multiplying Money

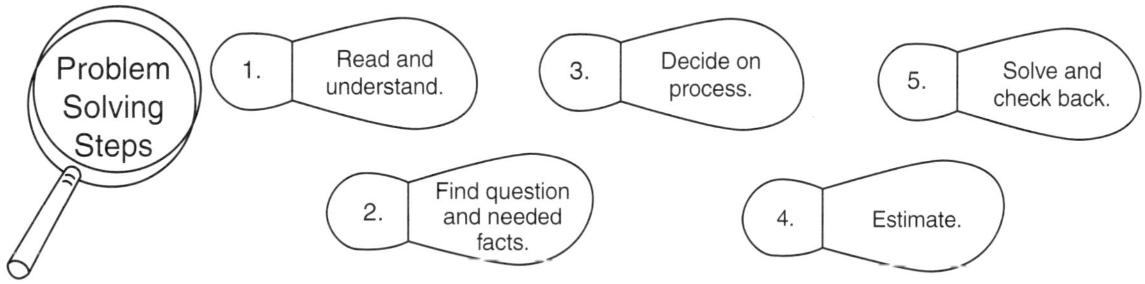

Willie bought 2 small notebooks at 29¢ each. How much did the notebooks cost?

Record:

$$\begin{array}{r} 29\text{¢} \\ \times\ \ 2 \\ \hline 58\text{¢} \end{array} \quad \text{or} \quad \begin{array}{r} \$0.29 \\ \times\ \ \ \ 2 \\ \hline \$0.58 \end{array}$$

Find the products. Write your answer in two ways, with a ¢ sign and a $ sign.

1. $\begin{array}{r} 35\text{¢} \\ \times\ 2 \\ \hline \end{array}$

2. $\begin{array}{r} 64\text{¢} \\ \times\ 5 \\ \hline \end{array}$

3. $\begin{array}{r} 125\text{¢} \\ \times\ 3 \\ \hline \end{array}$

4. $\begin{array}{r} 240\text{¢} \\ \times\ 6 \\ \hline \end{array}$

5. Ong bought 5 pencils at 12¢ each. How much did the pencils cost?

______

6. Nick bought 4 large notebooks at 85¢ apiece. How much did all the notebooks cost?

______

7. Amy bought 1 small notebook for 29¢ and 2 pencils for 12¢ apiece. How much did she spend?

______

8. Willie's grandfather bought 2 decks of cards for $1.85 per deck. He gave the clerk a $5.00 bill. How much change did he receive?

______

You finish the story. Ask a mathematical question and find the answer.

9. Nick and Willie went shopping in the music store. They found tapes on sale for $3.25 apiece.

______

10. Ong had $20.00. She found records on sale for $6.50 each.

______

Estimating the Product

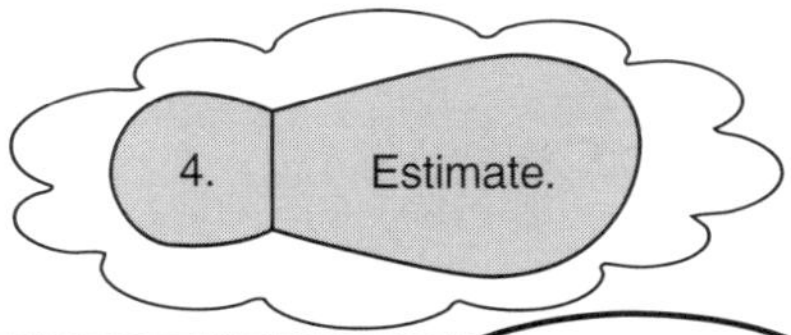

Ong wanted to buy 3 apples at 28¢ each. About how much money will she need?

To find about how much, I can round 28¢ to the nearest ten and multiply.

28 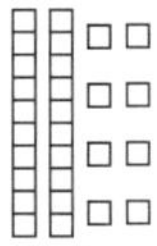 rounds to 30 

$3 \times 30¢ = 90¢$

1. Nick bought 4 bananas at 21¢ each. About how much did he spend?

   ____

2. Nicole bought 5 lemons at 19¢ each. Will $1.00 be enough to buy the lemons?

   ____

3. Each fourth grade class has 27 students. There are 3 classes of fourth graders. Estimate the number of fourth graders.

   ____

4. 74 fourth graders eat lunch in the school cafeteria every day. Estimate the number of lunches served during one week.

   ____

Estimate the products by rounding the larger number to the nearest ten.

	rounds to	estimate		rounds to	estimate
5. 3×49	___ × ___	= ______	**6.** 4×38	___ × ___	= ______
7. 6×21	___ × ___	= ______	**8.** 5×27	___ × ___	= ______
9. 7×83	___ × ___	= ______	**10.** 2×89	___ × ___	= ______
11. 8×17	___ × ___	= ______	**12.** 8×12	___ × ___	= ______
13. 65×7	___ × ___	= ______	**14.** 35×6	___ × ___	= ______
15. 88×4	___ × ___	= ______	**16.** 7×43	___ × ___	= ______
17. 2×93	___ × ___	= ______	**18.** 66×3	___ × ___	= ______
19. 76×8	___ × ___	= ______	**20.** 46×9	___ × ___	= ______

Problem Solving: Draw a Picture

Every multiplication fact can be modeled as an array. The sides of the array are the factors. The total number of square units is the product.

$2 \times 3 = 6$

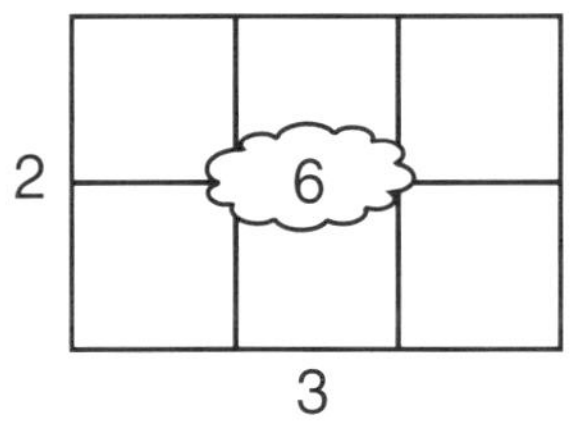

Multiplication of larger numbers is modeled in the same way.

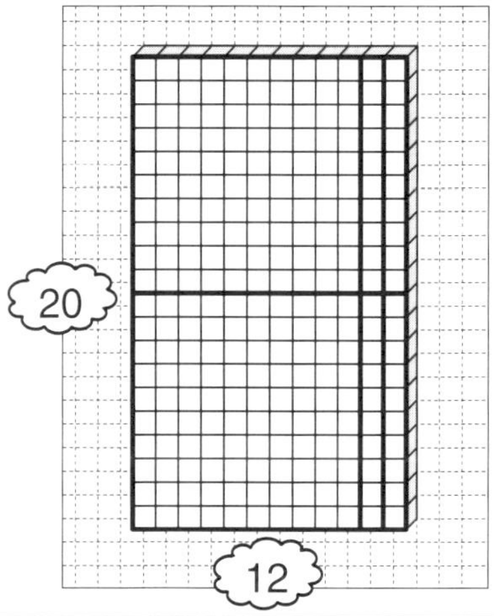

$$\begin{array}{r} 12 \\ \times\ 20 \\ \hline 240 \end{array}$$

Use base ten blocks on centimeter graph paper to find each product. Check with paper and pencil and a calculator.

1. $30 \times 20 =$ ______ **2.** $40 \times 50 =$ ______ **3.** $70 \times 30 =$ ______

4. $20 \times 14 =$ ______ **5.** $30 \times 12 =$ ______ **6.** $40 \times 21 =$ ______

7. $30 \times 30¢ =$ ______ **8.** $50 \times 80¢ =$ ______ **9.** $70 \times 50¢ =$ ______

10. $21 \times 20¢ =$ ______ **11.** $13 \times 30¢ =$ ______ **12.** $23 \times 20¢ =$ ______

13. $\begin{array}{r} 34 \\ \times\ 20 \\ \hline \end{array}$ **14.** $\begin{array}{r} 23 \\ \times\ 30 \\ \hline \end{array}$ **15.** $\begin{array}{r} 43 \\ \times\ 20 \\ \hline \end{array}$ **16.** $\begin{array}{r} 11 \\ \times\ 50 \\ \hline \end{array}$

17. $\begin{array}{r} 71 \\ \times\ 20 \\ \hline \end{array}$ **18.** $\begin{array}{r} 21 \\ \times\ 60 \\ \hline \end{array}$ **19.** $\begin{array}{r} 32 \\ \times\ 40 \\ \hline \end{array}$ **20.** $\begin{array}{r} 31 \\ \times\ 50 \\ \hline \end{array}$

21. A frozen juice bar costs 20¢. How much would 2 dozen frozen juice bars cost? ______

22. How many minutes are in 24 hours? ______

Problem Solving with Multiplication

The Greens went fishing. There were 5 people in the boat. Each person caught 7 fish. How many fish did they catch in all?

This problem puts together groups of equal size, so it is a multiplication problem.

$$\begin{array}{r} 7 \\ \times\ 5 \\ \hline 35 \end{array}$$

1. Tony drank 6 glasses of milk every day for 7 days. How many glasses of milk did he drink in all?

   ____

2. Erica helped her dad chop 6 trees into logs. They cut each tree into 9 logs. How many logs did they chop altogether?

   ____

3. Mary went to summer school four days a week for four weeks. How many days did she go to school in all?

   ____

4. The radio station played Jennie's favorite song 4 times on Monday and 9 times on Tuesday. How many times did they play the song in all?

   ____

5. Mr. Teller filled 8 shelves with 9 boxes of "Wowee" cereal on each shelf. How many boxes of cereal were on all the shelves?

   ____

6. Charlie Chipmunk dug 5 holes in the woods. In each hole he placed 5 acorns. How many acorns did he hide in all?

   ____

7. The Fridley track team has 8 members. Each member runs 7 miles a week. How many miles does the team run in a week?

   ____

8. The softball team had 3 practices a week for 8 weeks. How many practices did they have in all?

   ____

9. Kurt collects baseball hats. He put 6 hats on each of the 3 shelves in his bedroom. How many hats are on Kurt's shelves?

   ____

10. Miss Hall walks 4 flights of stairs to get to her office. Each flight has 7 steps on it. How many steps does she walk to get to her office?

    ____

Division: Sharing Groups of Equal Size

Share cookies.

	Number of Cookies	Number of Plates	How many cookies on each plate?
1.	12	6	
2.	9	3	
3.	12	4	
4.	10	5	
5.	12	3	
6.	14	2	
7.	3	3	
8.	15	5	
9.	16	4	
10.	20	5	

Division and Multiplication as Opposites

Multiplication puts together groups of equal size.

2 groups of 5 is 10.

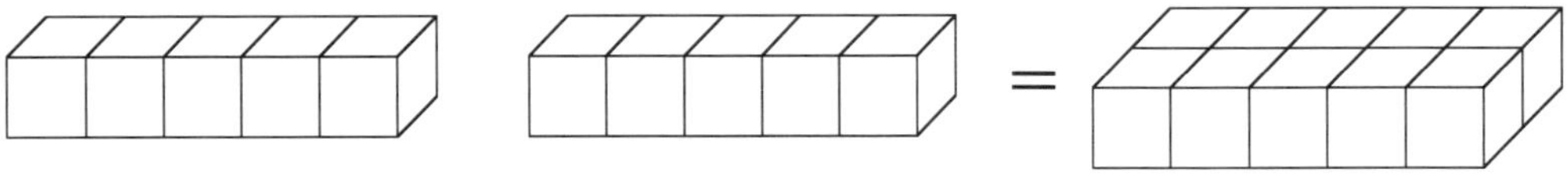

$$\begin{array}{r}5\\ \times 2\\ \hline 10\end{array}$$

Division takes apart groups of equal size.

10 divided into 2 groups is 5.

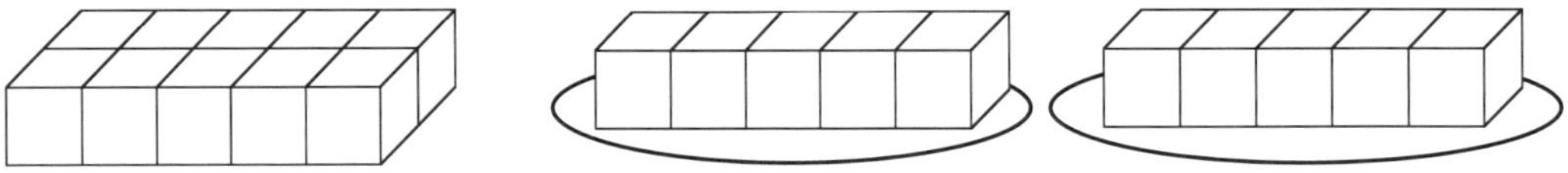

$$\begin{array}{r}5\\ 2\overline{)10}\end{array}$$

Write two division facts for each multiplication fact.

1. $2 \times 4 = 8$

$\begin{array}{r}4\\ 2\overline{)8}\end{array}$

$\begin{array}{r}2\\ 4\overline{)8}\end{array}$

2. $5 \times 3 = 15$

$\overline{)}$

$\overline{)}$

3. $4 \times 3 = 12$

$\overline{)}$

$\overline{)}$

4. $2 \times 9 = 18$

$\overline{)}$

$\overline{)}$

5. $3 \times 6 = 18$

$\overline{)}$

$\overline{)}$

6. $8 \times 2 = 16$

$\overline{)}$

$\overline{)}$

7. $4 \times 7 = 28$

$\overline{)}$

$\overline{)}$

8. $9 \times 3 = 27$

$\overline{)}$

$\overline{)}$

9. 12  6 

How many  for each  ?

____

10. 8 

2  on each 

How many  ?

____

Relating Multiplication and Division

Use blocks or squares to find the missing side. Write a division fact.

1. Use 30 blocks with 5 units on one side. Missing side = _____ Division Fact: __________________	**2.** Use 12 blocks with 2 units on one side. Missing side = _____ Division Fact: __________________
3. Use 12 blocks with 3 units on one side. Missing side = _____ Division Fact: __________________	**4.** Use 12 blocks with 4 units on one side. Missing side = _____ Division Fact: __________________

How many units on the missing side? Write the division fact in two ways.

5.

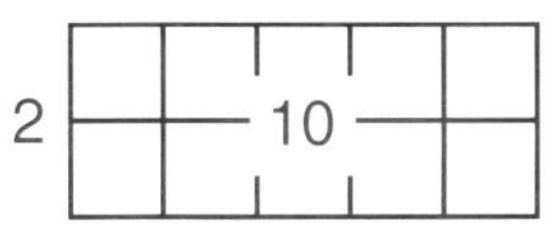

? = _______

__________ and __________

6.

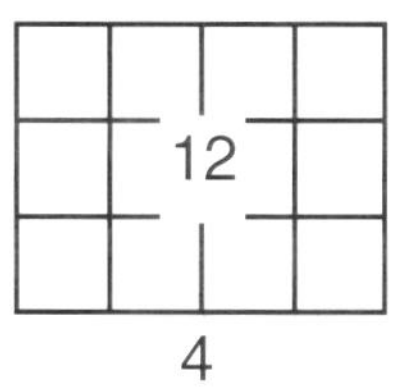

? = _______

__________ and __________

2-Digit Quotients

A class of 36 students is going on a field trip in 3 vans. How many students will ride in each van?

$3\overline{)36}$

Build 36 (36 = 3 tens and 6 ones). Share tens and ones among 3 plates.

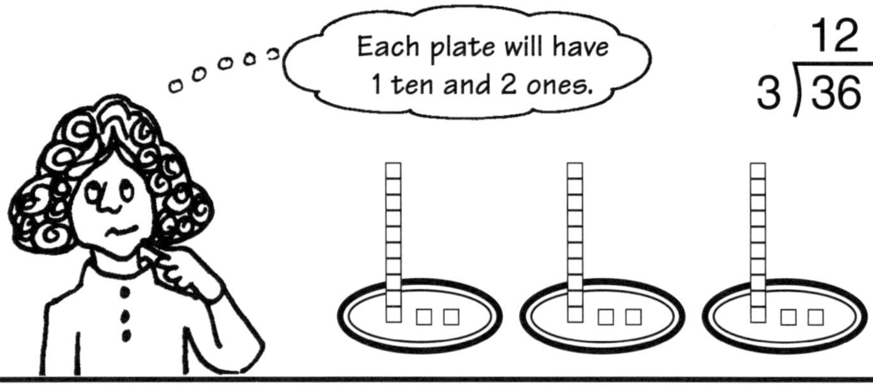

$$\begin{array}{r} 12 \\ 3\overline{)36} \end{array}$$

Build and share base ten blocks.

1. $2\overline{)40}$	**2.** $2\overline{)42}$	**3.** $2\overline{)68}$	**4.** $2\overline{)46}$
5. $2\overline{)66}$	**6.** $5\overline{)50}$	**7.** $5\overline{)55}$	**8.** $3\overline{)60}$
9. $3\overline{)63}$	**10.** $3\overline{)36}$	**11.** $3\overline{)96}$	**12.** $3\overline{)60}$
13. $4\overline{)84}$	**14.** $4\overline{)80}$	**15.** $3\overline{)63}$	**16.** $3\overline{)39}$

17. 8 cakes were to be shared equally among 80 people. How many pieces must each cake be cut into?

____

18. There were 96 pencils. 3 were put in each bag. How many bags were used?

____

Using a Model to Discover Division Patterns

A class of 36 students is going on a field trip in 2 buses. How many students will be on each bus?

$2\overline{)36}$

Build 36 and share tens and ones.

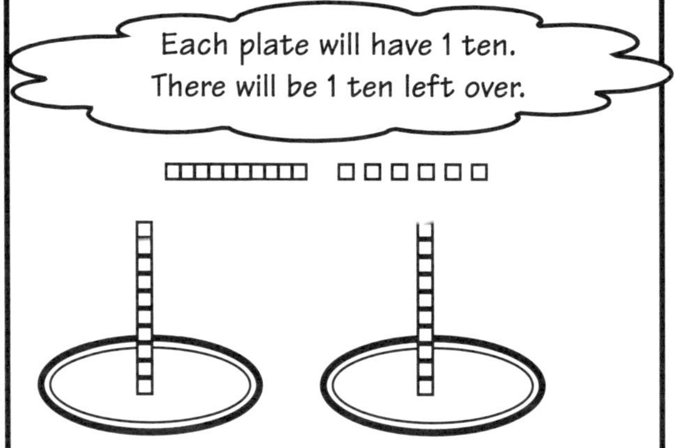

Exchange 1 ten for 10 ones.
Share 16 ones.

$$\begin{array}{r}18\\2\overline{)36}\end{array}$$

Build and share base ten blocks. Record the answer.

1. $4\overline{)52}$ **2.** $5\overline{)75}$ **3.** $3\overline{)78}$ **4.** $7\overline{)84}$

5. $3\overline{)75}$ **6.** $4\overline{)96}$ **7.** $7\overline{)91}$ **8.** $8\overline{)96}$

9. $2\overline{)58}$ **10.** $2\overline{)52}$ **11.** $3\overline{)72}$ **12.** $6\overline{)78}$

Dividing by 1 Digit with Regrouping

Marian and Kim babysit together and share their earnings at the end of each month. During January, the girls received $136 for babysitting. What is each girl's share?

This problem separates groups of equal size.

$2\overline{)136}$

Build 136.
Share hundreds.

1 hundred cannot be shared. After exchanging 1 hundred for 10 tens, there are 13 tens.

Share tens.

Each plate receives 6 tens. 1 ten is exchanged for 10 ones, so there are 16 ones.

Share ones.

Each plate receives 8 ones.

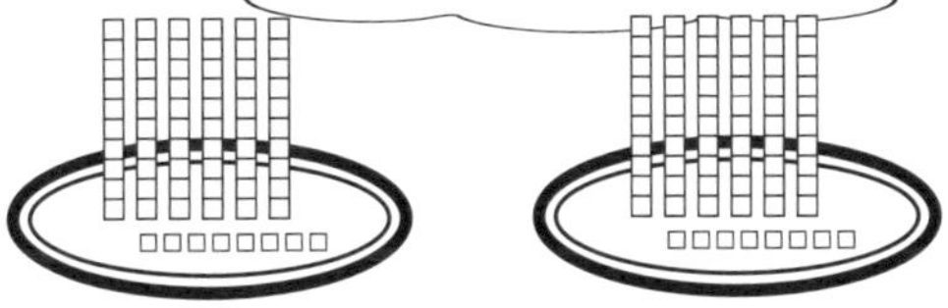

Use the division pattern **DMS↓** to record.

1. Share the hundreds.

$$\begin{array}{r} 0 \\ 2\overline{)136} \\ \underline{0} \\ 13 \end{array}$$

2. Share the tens.

$$\begin{array}{r} 06 \\ 2\overline{)136} \\ \underline{0} \\ 13 \\ \underline{12} \\ 1 \end{array}$$

3. Share the ones.

$$\begin{array}{r} 068 \\ 2\overline{)136} \\ \underline{0} \\ 13 \\ \underline{12} \\ 16 \\ \underline{16} \end{array}$$

Use base ten blocks to divide. Use the pattern to record.

1. $2\overline{)126}$ **2.** $2\overline{)104}$ **3.** $2\overline{)168}$ **4.** $2\overline{)140}$

5. $3\overline{)129}$ **6.** $3\overline{)246}$ **7.** $3\overline{)153}$ **8.** $3\overline{)273}$

9. $4\overline{)164}$ **10.** $4\overline{)208}$ **11.** $5\overline{)105}$ **12.** $4\overline{)324}$

4-Digit Quotients

Mr. and Mrs. Traugot drove 1346 miles in 2 days. If they drove an equal number of miles each day, how many miles did they drive per day?

$2\overline{)1346}$

```
   0673
2 )1346
   12
    14
    14
      6
      6
```

Build 1346.

Share thousands. — 1 thousand must be exchanged for 10 hundreds making a total of 13 hundreds.

Share hundreds. — There will be 6 hundreds on each plate. After 1 hundred is exchanged for 10 tens, there are 14 tens.

Share tens. — There will be 7 tens on each plate.

Share ones. — There will be 3 ones on each plate.

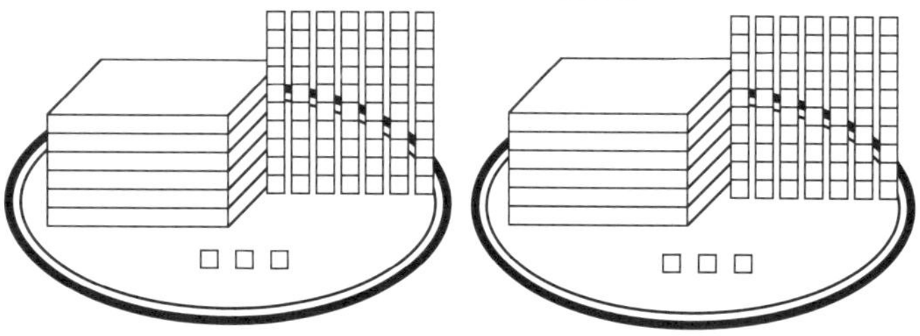

6 hundreds 7 tens 3 ones = 673

Use base ten blocks to find the quotients.

1. $4\overline{)1820}$ **2.** $5\overline{)1185}$ **3.** $2\overline{)1534}$

Use the division pattern (DMS↓) to find the quotients.

4. $6\overline{)8244}$ **5.** $7\overline{)9184}$ **6.** $8\overline{)9232}$

7. $2\overline{)2714}$ **8.** $4\overline{)8444}$ **9.** $5\overline{)5585}$

Problem Solving Steps

1. Read and understand.
2. Find question and needed facts.
3. Decide on process.
4. Estimate.
5. Solve and check back.

There are 4 buses with the same number of students on each bus. There are 144 students in all. How many are on each bus?

This problem takes apart equal size groups. It is a division problem.

$$\begin{array}{r} 36 \\ 4\overline{)144} \\ \underline{12} \\ 24 \\ \underline{24} \end{array}$$

36 students on each bus

1. The cast of the school play has 210 pages of lines to learn. The play is in 7 weeks. How many pages should they learn each week? ______

2. The school cafeteria served 936 pieces of pizza on Monday. If each student received 3 pieces, how many students did they serve? ______

3. Matt has taped 427 hours of television. If each tape is 7 hours long, how many tapes does he have? ______

4. The Brown family has 756 plants in their garden. With 9 plants in a row, how many rows are there in their garden? ______

5. The highway crew fixed 256 miles of road at the rate of 8 miles per day. How many days did they work? ______

6. A farmer wants to collect 282 eggs each week. If one hen can lay 6 eggs a week, how many hens should the farmer buy? ______

7. Ray packed 160 crayons in boxes. If he put 8 crayons in each box, how many boxes did he fill? ______

8. Baby Bonnie drank 200 ounces of milk in a week. If each bottle holds 8 ounces, how many bottles did she drink? ______

9. Mom had 27 apples. 8 apples are needed for a pie. How many pies can she make? How many apples are left? ______

10. The zookeeper had 86 fish. He will give each seal 3 fish. How many seals can he feed? How many fish will be left? ______

Using Models to Find Fractions

Write the fraction and word name for each shaded part.

1.

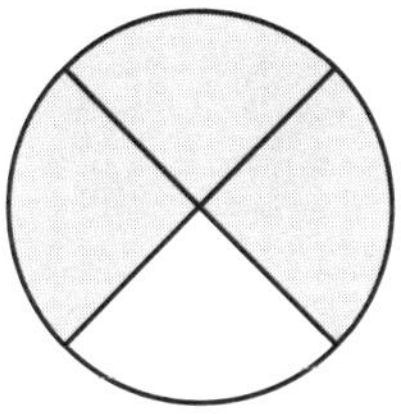

$\frac{3}{4}$

three-fourths

2.

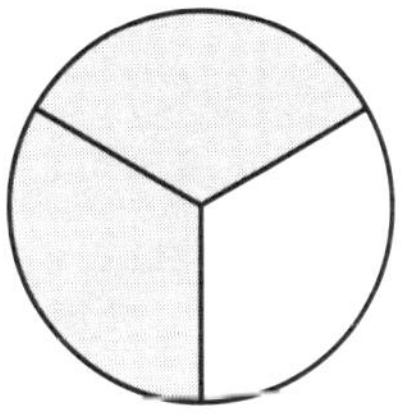

3.

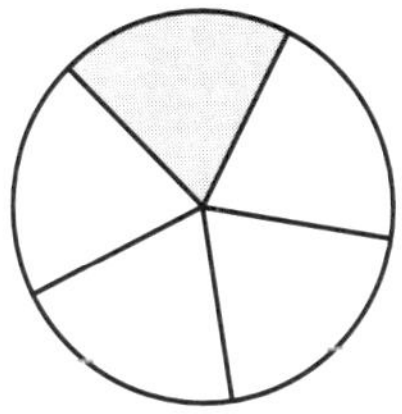

4.

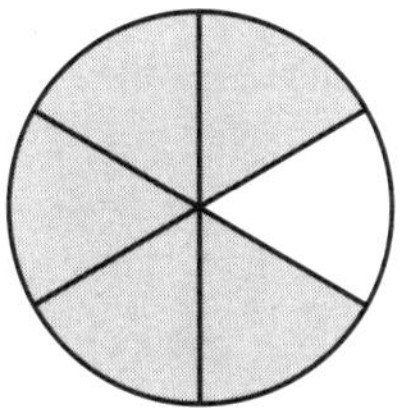

5.

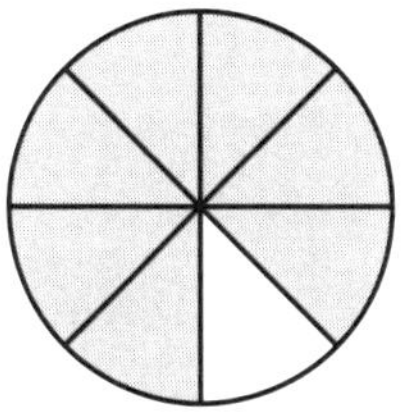

6.

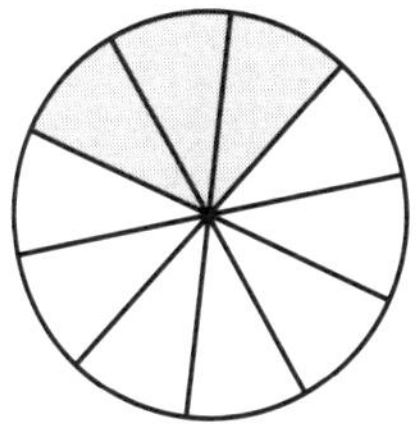

7.

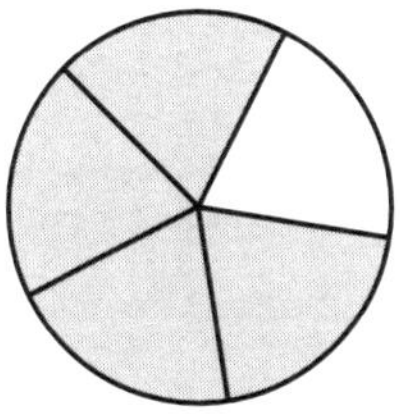

8.

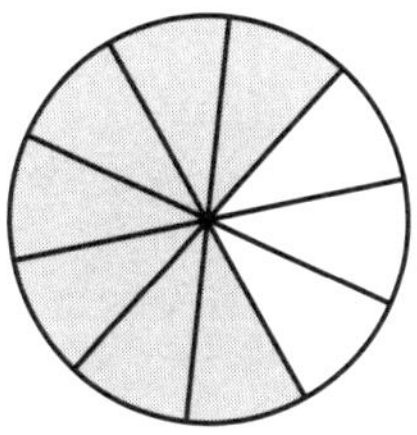

9.

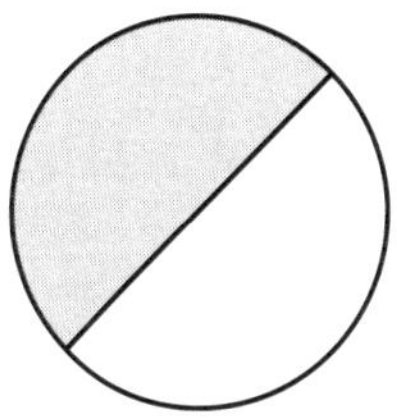

Using Models to Find Fractions

Here is a set of children.

1. How many children in the total set? __________

2. How many children are wearing glasses? __________

3. What fraction of the children are wearing glasses? __________

4. What fraction of the children are not wearing glasses? __________

What fraction of each set is shaded?

5. 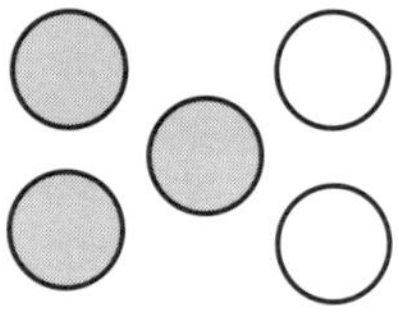

____

6. 

____

7. 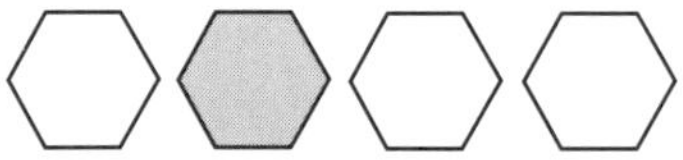

____

8. 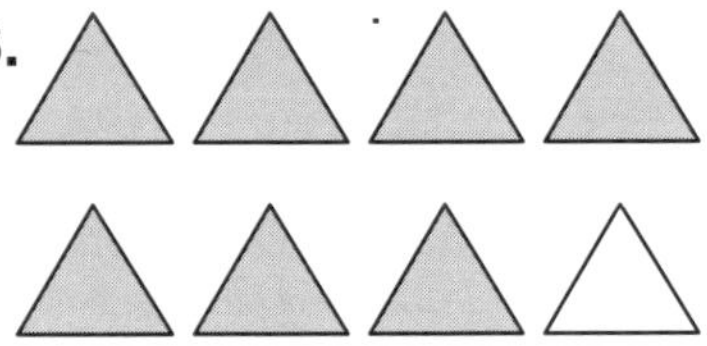

____

9. 

____

10. 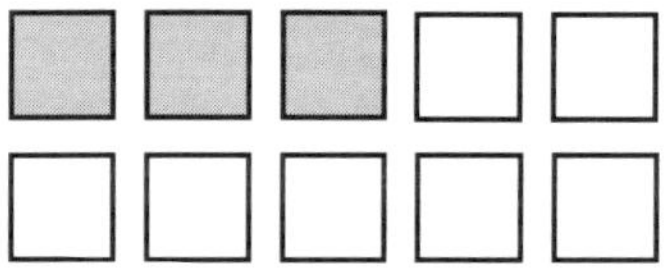

____

Shade a part of the set for each fraction.

11. $\frac{3}{4}$ 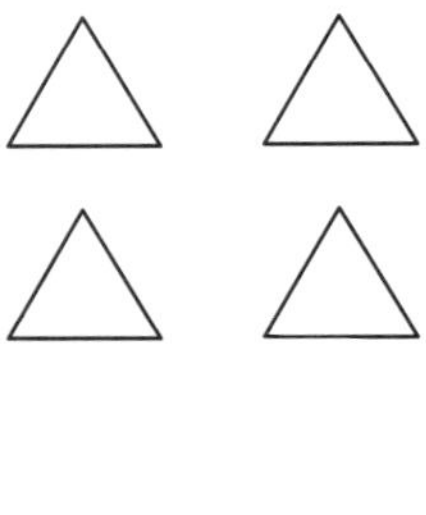

12. $\frac{5}{8}$ 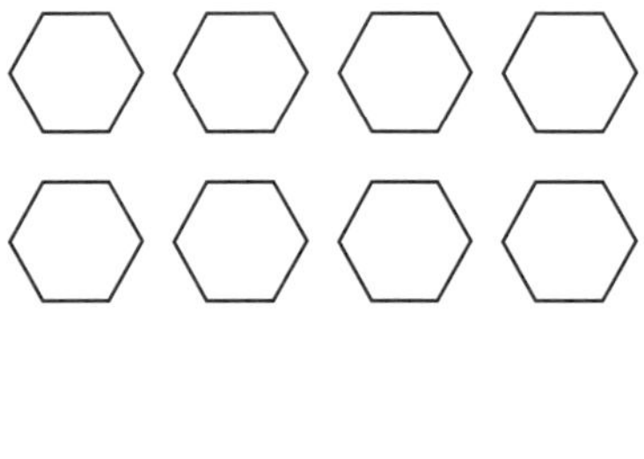

Comparing Fractions with the Same Number of Matching Parts

What symbol (>, <, =) goes in the box?

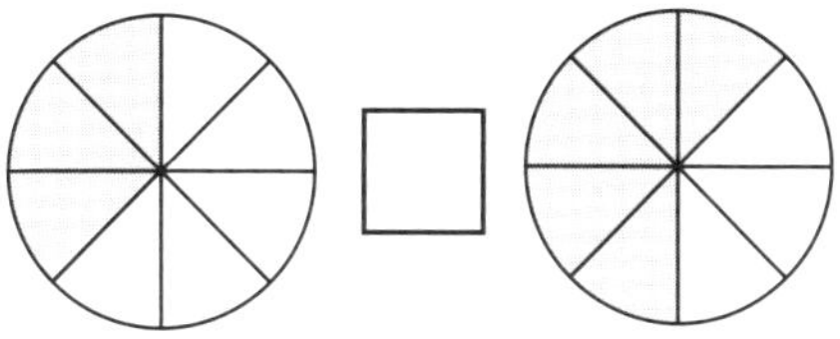

$\frac{3}{8}$ < $\frac{5}{8}$

What symbol (>, <, =) goes in the box?

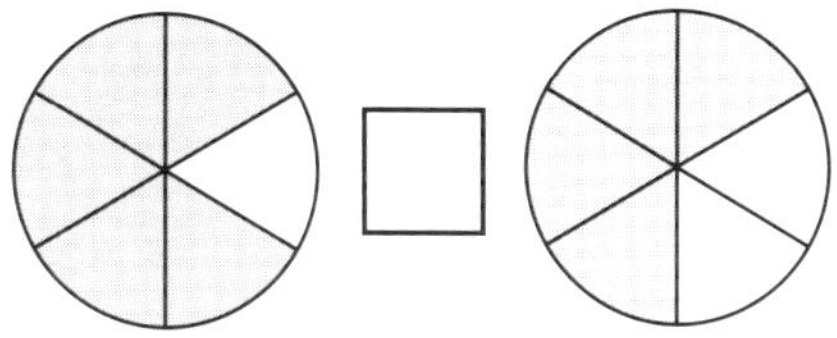

$\frac{5}{6}$ > $\frac{4}{6}$

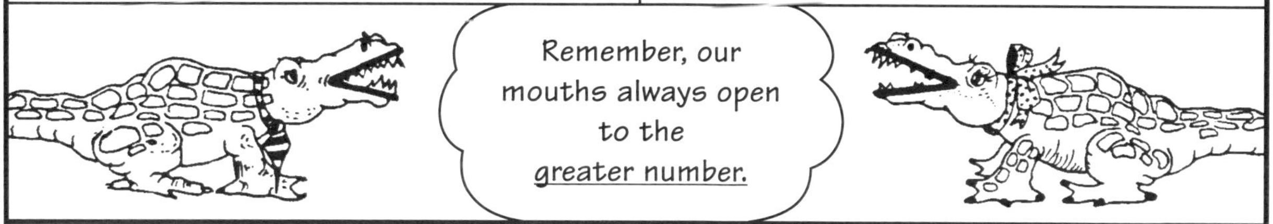

$\frac{1}{4}$ 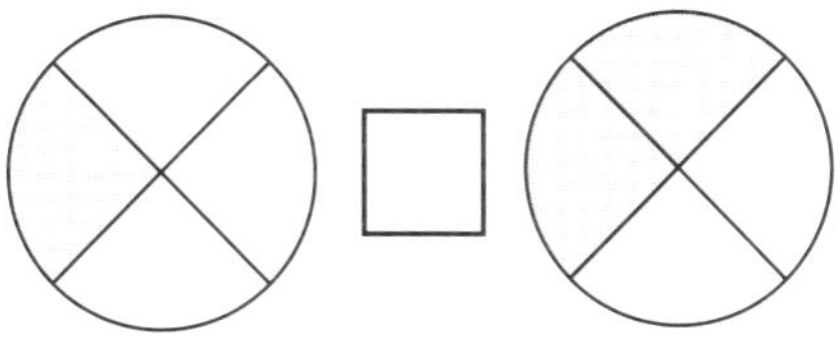 $\frac{2}{4}$

1. > , < , = ? ________

$\frac{3}{3}$ 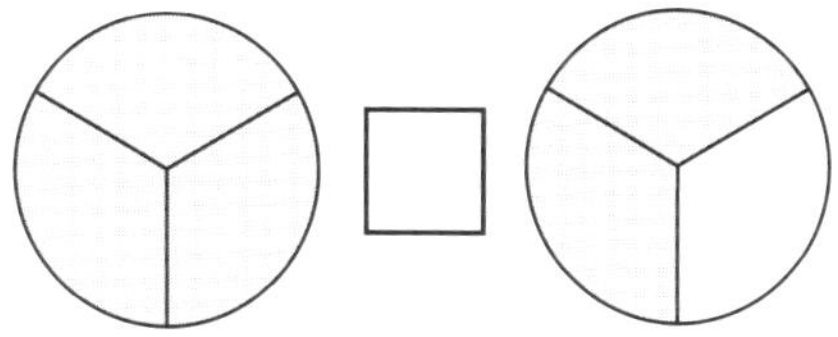 $\frac{2}{3}$

2. > , < , = ? ________

$\frac{2}{5}$ 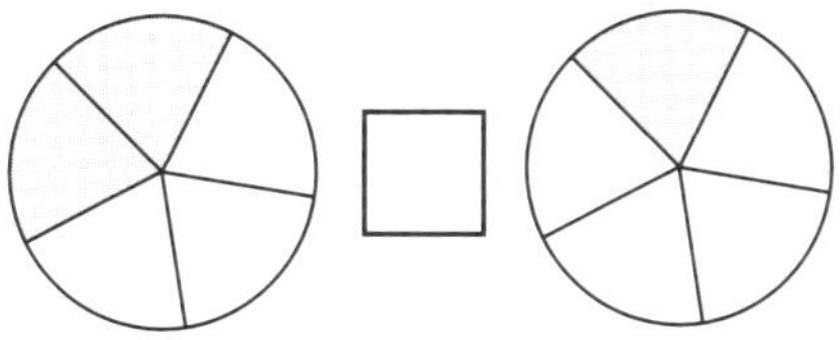 $\frac{1}{5}$

3. > , < , = ? ________

$\frac{5}{10}$ 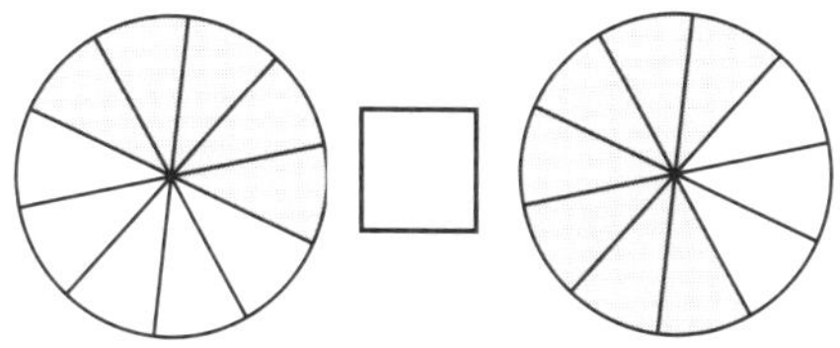 $\frac{7}{10}$

4. > , < , = ? ________

$\frac{1}{2}$ 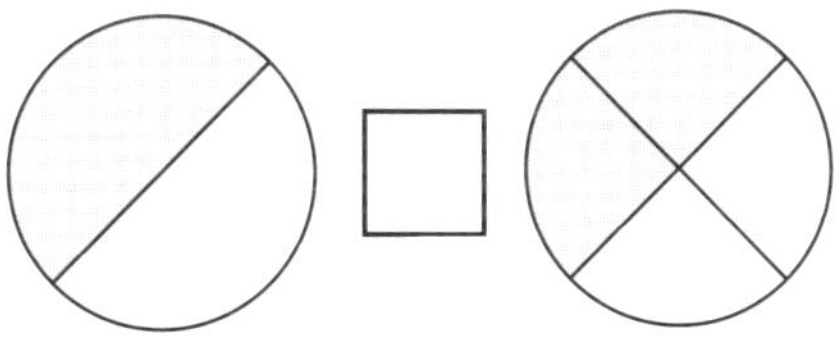 $\frac{2}{4}$

5. > , < , = ? ________

$\frac{3}{6}$ 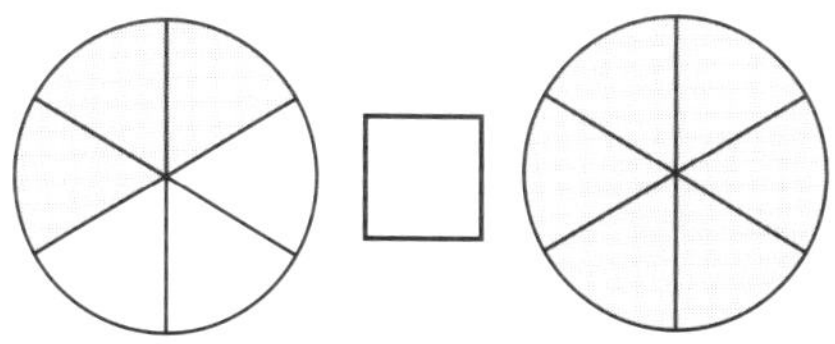 $\frac{6}{6}$

6. > , < , = ? ________

$\frac{7}{8}$ 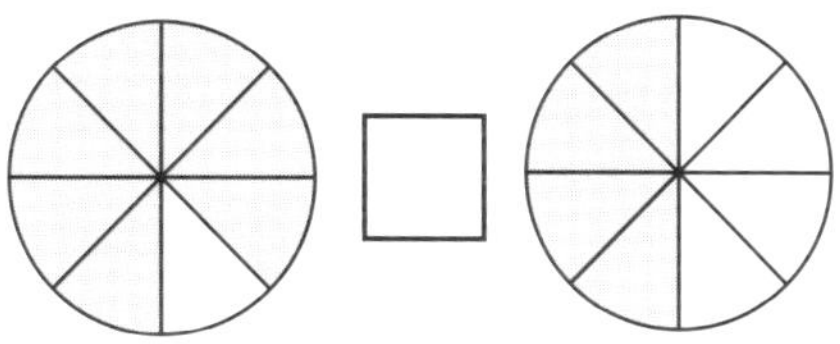 $\frac{4}{8}$

7. > , < , = ? ________

$\frac{3}{6}$ 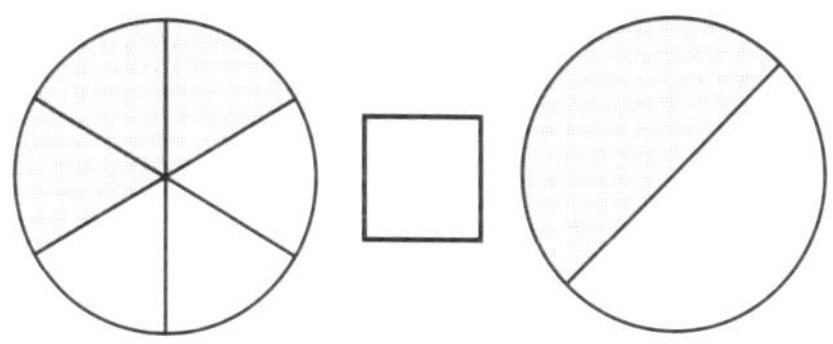 $\frac{1}{2}$

8. > , < , = ? ________

$\frac{4}{8}$ 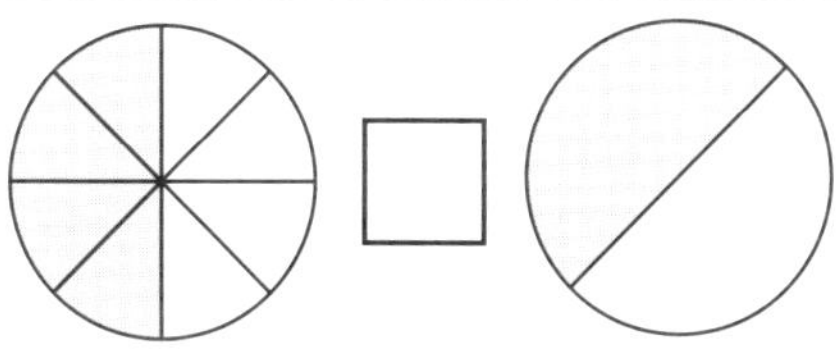 $\frac{1}{2}$

9. > , < , = ? ________

$\frac{3}{8}$ 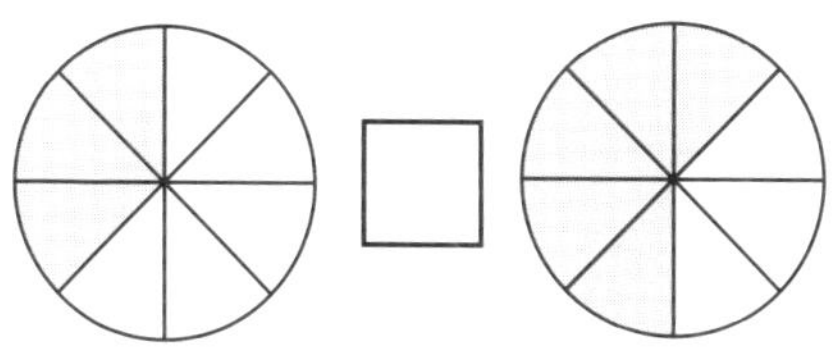 $\frac{5}{8}$

10. > , < , = ? ________

Other Names for One

Each of these circles has all of its parts shaded.
Each of these circles are other names for 1.

$1 = \frac{2}{2} = 1$ $\quad$ $1 = \frac{4}{4} = 1$ $\quad$ $1 = \frac{8}{8} = 1$ $\quad$ $1 = \frac{6}{6} = 1$

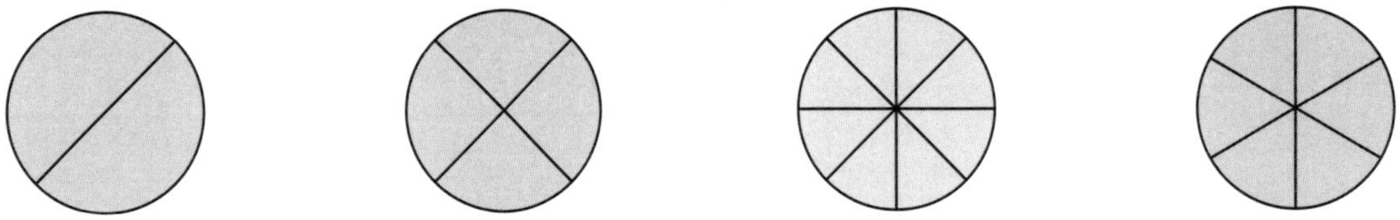

The simplest name for all these fractions is 1.

1. Count by $\frac{1}{5}$'s from $\frac{1}{5}$ to $\frac{5}{5}$ or 1.

_____, _____, _____, _____, _____, or _____ .

2. Count by $\frac{1}{8}$'s from $\frac{1}{8}$ to $\frac{8}{8}$ or 1.

_____, _____, _____, _____, _____, _____, _____, _____, or _____ .

3. Count by $\frac{1}{6}$'s from $\frac{1}{6}$ to $\frac{6}{6}$ or 1.

_____, _____, _____, _____, _____, _____, or _____ .

4. Count by $\frac{1}{10}$'s from $\frac{1}{10}$ to $\frac{10}{10}$ or 1.

_____, _____, _____, _____, _____, _____, _____, _____, _____, _____, or _____ .

5. Which of these are other names for 1?

a. $\frac{7}{7}$ **b.** $\frac{1}{7}$ **c.** $\frac{8}{9}$ **d.** $\frac{5}{5}$ **e.** $\frac{12}{12}$ **f.** $\frac{1}{1}$

g. $\frac{1}{4}$ **h.** $\frac{10}{10}$ **i.** $\frac{2}{2}$ **j.** $\frac{5}{1}$ **k.** $\frac{3}{3}$ **l.** $\frac{12}{1}$

A. How many different fraction names can you write for the number 1?

B. Can you find a rule to help decide whether a fraction is equal to 1?

Adding Fractions

$$\frac{1}{4} + \frac{2}{4} = ?$$

Step 1. Build each of the fractions to be added.

$\frac{1}{4}$ + $\frac{2}{4}$

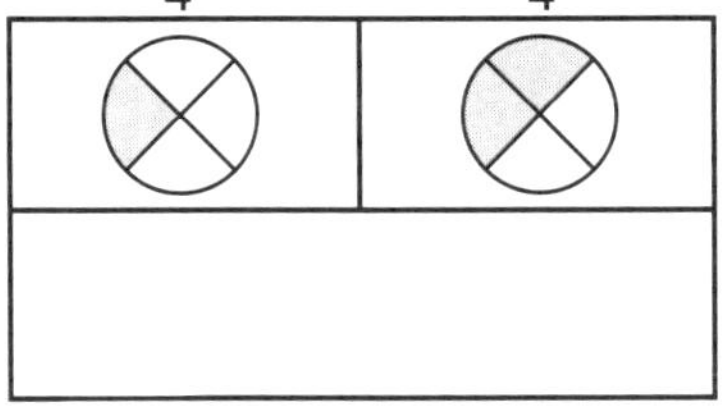

Step 2. Put the fractions together to show addition

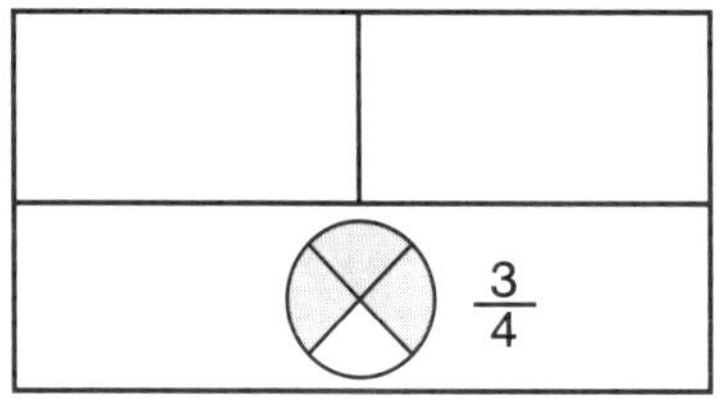

Step 3. Say the problem: "One-fourth plus two-fourths equals three-fourths."

1. $\frac{1}{4} + \frac{2}{4}$

2. $\frac{1}{8} + \frac{2}{8}$

3. $\frac{1}{5} + \frac{1}{5}$

4. $\frac{1}{10} + \frac{2}{10}$

5. $\frac{1}{3} + \frac{1}{3}$

6. $\frac{3}{8} + \frac{2}{8}$

7. $\frac{5}{10} + \frac{2}{10}$

8. $\frac{3}{5} + \frac{1}{5}$

9. $\frac{2}{5} + \frac{1}{5}$

10. $\frac{4}{8} + \frac{1}{8}$

11. $\frac{2}{4} + \frac{1}{4}$

12. $\frac{7}{10} + \frac{2}{10}$

13. $\frac{4}{10} + \frac{3}{10}$

14. $\frac{2}{8} + \frac{3}{8}$

15. $\frac{6}{8} + \frac{1}{8}$

16. $\frac{6}{10} + \frac{3}{10}$

Subtracting Fractions

The pattern for subtracting fractions divided into the same number of matching parts: write the difference (–) of the top numbers over the bottom number.	The bottom number stays the same. $\begin{array}{r} \frac{7}{8} \\ -\ \frac{2}{8} \\ \hline \frac{5}{8} \end{array}$ $7 - 2 = 5$ Subtract the top numbers

1. $\begin{array}{r} \frac{3}{4} \\ -\ \frac{2}{4} \\ \hline \end{array}$
2. $\begin{array}{r} \frac{5}{8} \\ -\ \frac{2}{8} \\ \hline \end{array}$
3. $\begin{array}{r} \frac{2}{3} \\ -\ \frac{1}{3} \\ \hline \end{array}$
4. $\begin{array}{r} \frac{3}{5} \\ -\ \frac{2}{5} \\ \hline \end{array}$
5. $\begin{array}{r} \frac{6}{8} \\ -\ \frac{3}{8} \\ \hline \end{array}$
6. $\begin{array}{r} \frac{3}{10} \\ -\ \frac{2}{10} \\ \hline \end{array}$
7. $\begin{array}{r} \frac{2}{4} \\ -\ \frac{1}{4} \\ \hline \end{array}$
8. $\begin{array}{r} \frac{4}{8} \\ -\ \frac{1}{8} \\ \hline \end{array}$
9. $\begin{array}{r} \frac{5}{10} \\ -\ \frac{4}{10} \\ \hline \end{array}$
10. $\begin{array}{r} \frac{4}{4} \\ -\ \frac{3}{4} \\ \hline \end{array}$
11. $\begin{array}{r} \frac{7}{8} \\ -\ \frac{3}{8} \\ \hline \end{array}$
12. $\begin{array}{r} \frac{2}{2} \\ -\ \frac{1}{2} \\ \hline \end{array}$
13. $\begin{array}{r} \frac{5}{10} \\ -\ \frac{2}{10} \\ \hline \end{array}$
14. $\begin{array}{r} \frac{3}{3} \\ -\ \frac{2}{3} \\ \hline \end{array}$
15. $\begin{array}{r} \frac{4}{5} \\ -\ \frac{3}{5} \\ \hline \end{array}$
16. $\begin{array}{r} \frac{7}{10} \\ -\ \frac{4}{10} \\ \hline \end{array}$

Sneaky Pete found these errors. Can you find what is wrong?

1. $\begin{array}{r} \frac{3}{4} \\ -\ \frac{2}{4} \\ \hline \frac{1}{0} \end{array}$
2. $\begin{array}{r} \frac{5}{8} \\ -\ \frac{2}{8} \\ \hline \frac{3}{0} \end{array}$
3. $\begin{array}{r} \frac{2}{3} \\ -\ \frac{1}{3} \\ \hline \frac{1}{0} \end{array}$
4. $\begin{array}{r} \frac{3}{5} \\ -\ \frac{2}{5} \\ \hline \frac{1}{0} \end{array}$

Lines and Line Segments

Here is a picture of a **line.** A line never stops.
The arrows at the end mean that it goes on forever.

Part of this line is marked off by the endpoints B and C.
The part of the line between B and C is called a **line segment.**

1. What is the name of this figure? ______

A line **B** line segment **C** angle **D** ray

2. What is the name of this figure? ______

A line **B** line segment **C** angle **D** ray

3. Which is a line segment? ______

A

C 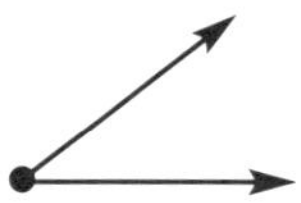

B 

D

4. Which is a line? ______

A 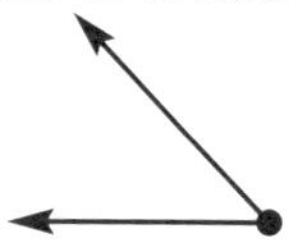

C 

B 

D 

5. Draw and label line segment AB.

6. Draw a line through point P.

7.

What is the name?

____

8.

What is the name?

____

Rays and Angles

A **ray** is part of a line.

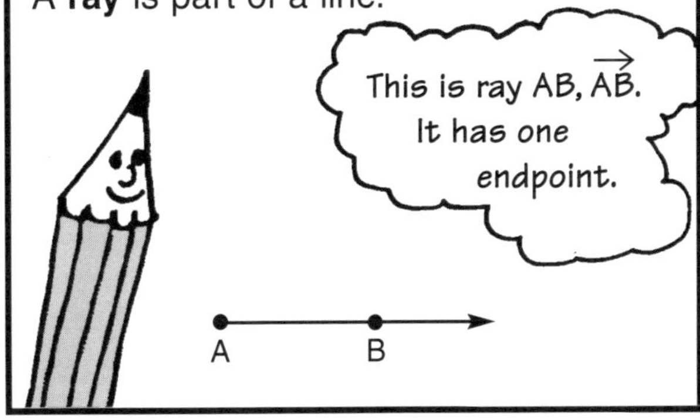

An **angle** is two rays having the same endpoint. The common endpoint is the **vertex**.

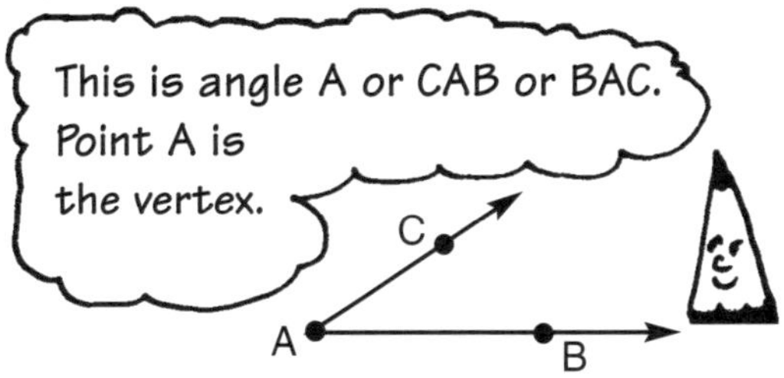

Complete.

1. 

What is this figure called?
How many endpoints does it have?

______________________ _____

2. 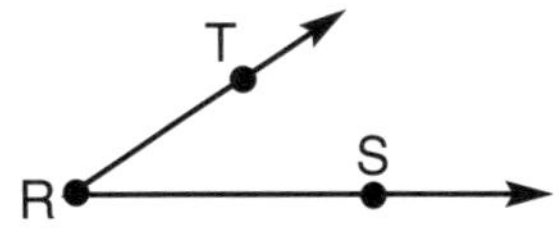

What is this figure called?
What is the vertex?

______________________ _____

3. Draw a figure showing $\overrightarrow{CD}$.

4. Draw a figure showing $\overrightarrow{CD}$ and $\overrightarrow{CE}$ having a common endpoint C.

5. Name this figure: ______________

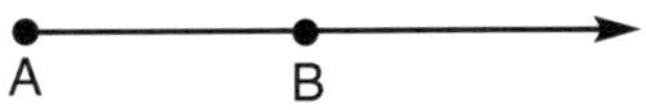

6. Name this figure in three ways:

________ ________ ________

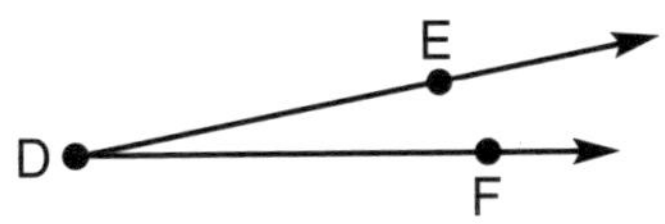

7. Which of the following are rays?

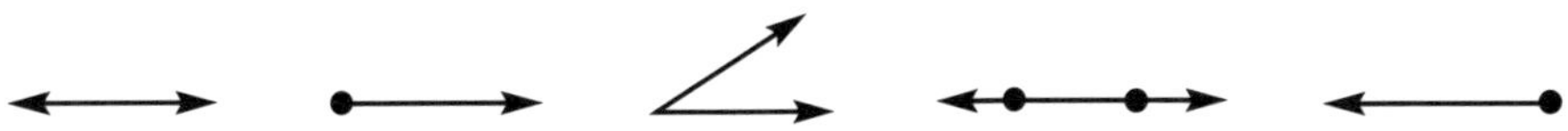

A **B** **C** **D** **E**

8. Which of the following are angles?

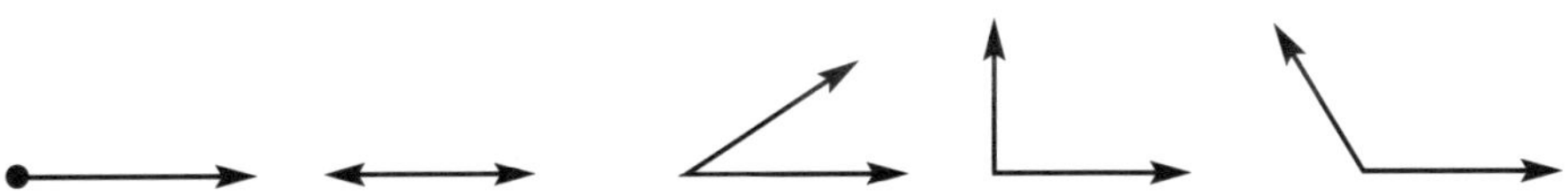

A **B** **C** **D** **E**

Perpendicular Lines

Intersecting lines which meet at square angles are called perpendicular lines.

The corners of many rooms and most picture frames are made of perpendicular lines.

These lines are perpendicular to each other.

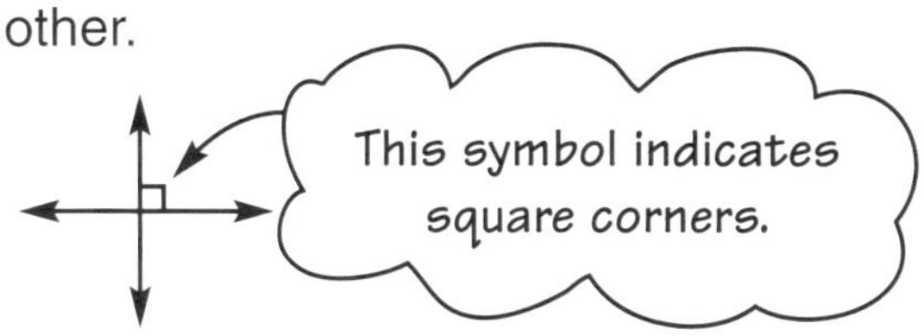

1. Which lines are perpendicular? ______

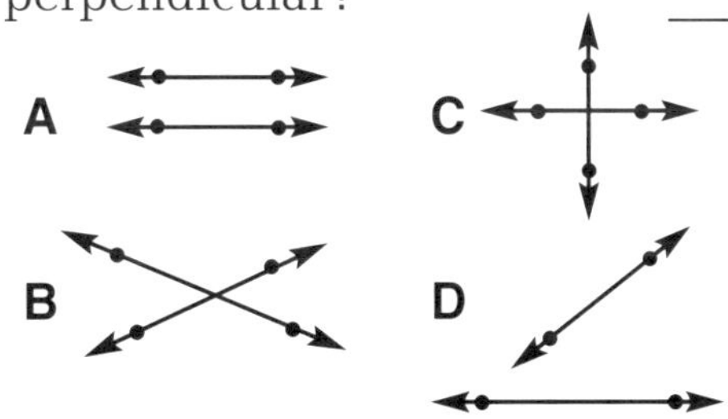

2. Which lines meet at square corners? ______

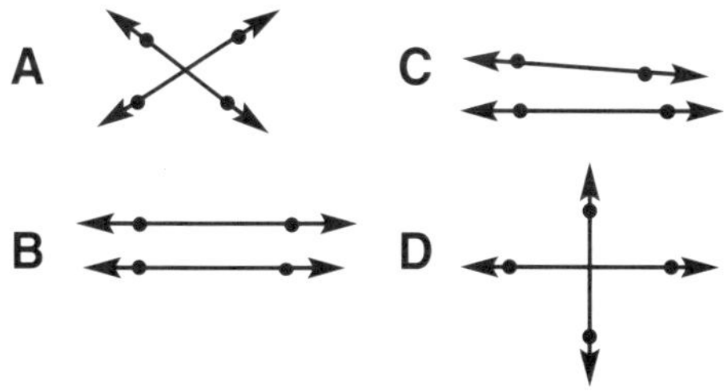

3. Draw a line perpendicular to line AB.

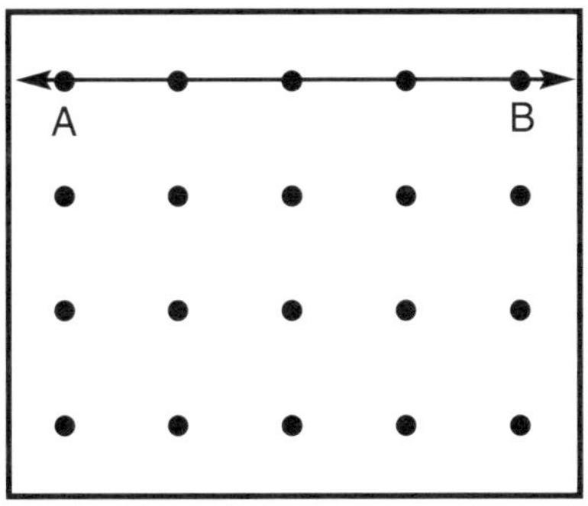

4. Draw a line perpendicular to line AB.

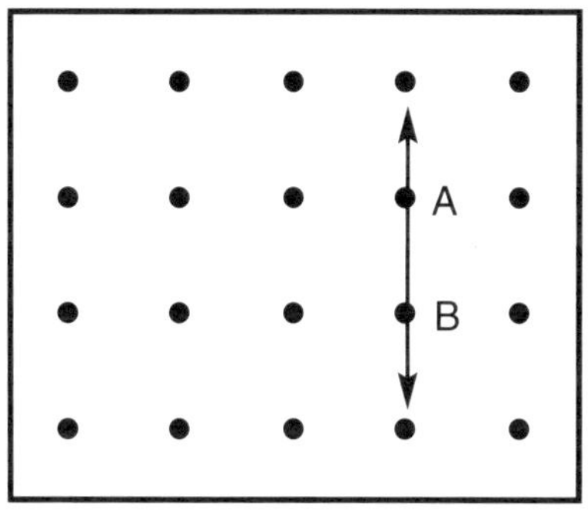

5. Draw a line perpendicular to line XY.

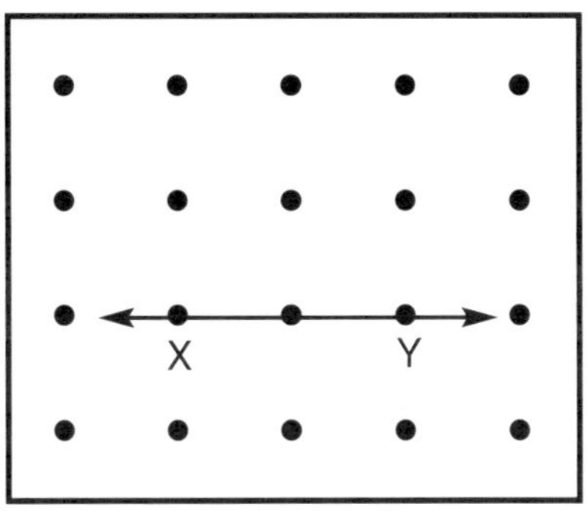

6. Draw a line perpendicular to line XY.

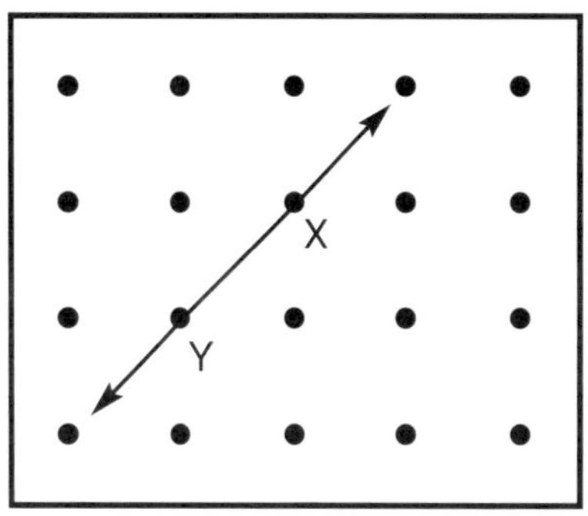

Find examples of perpendicular lines about you.

Congruent Figures

Two figures are congruent if they have the same shape and size.
They will match when one is placed over the other.

Which figure is congruent to the shaded figure in each row?

1.

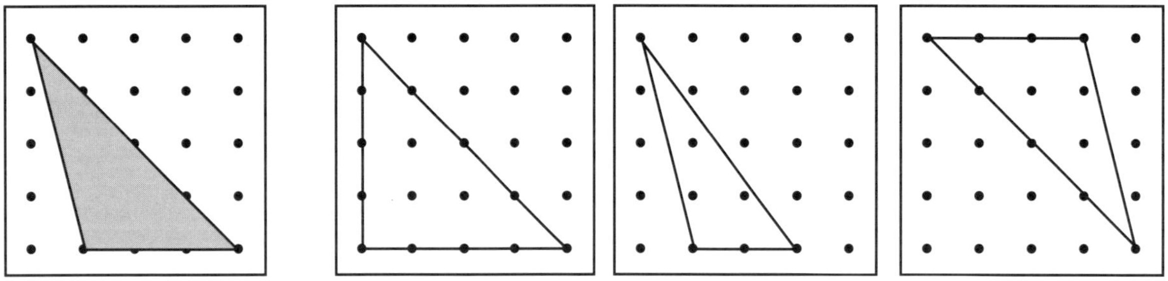

A B C

2.

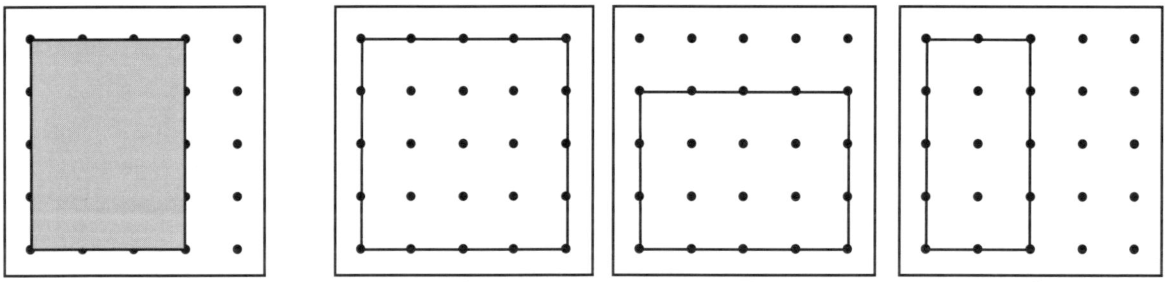

A B C

3.

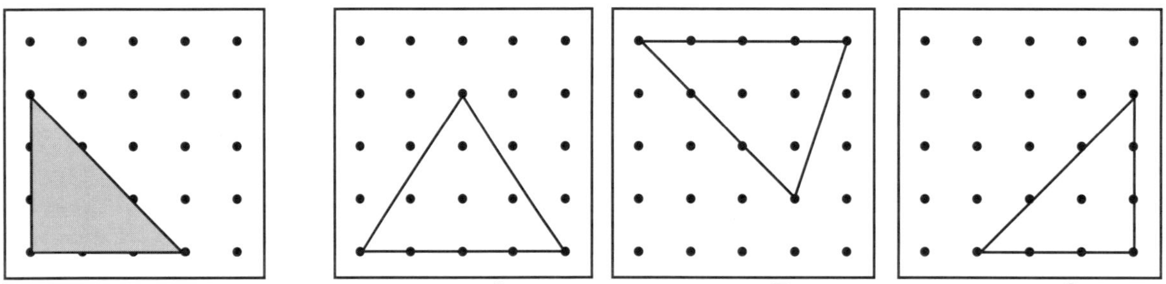

A B C

Teacher Note: Have students work together in pairs with geoboards. One student makes a figure with a rubber band. The other student makes a figure congruent to the first figure.

Telling Time to 15 Minutes

Each clock shows that the time is 15 minutes after 8.	Each clock shows that the time is 45 minutes after 1.
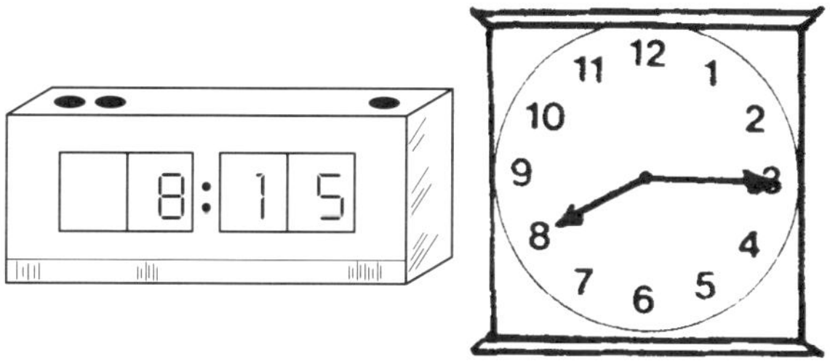	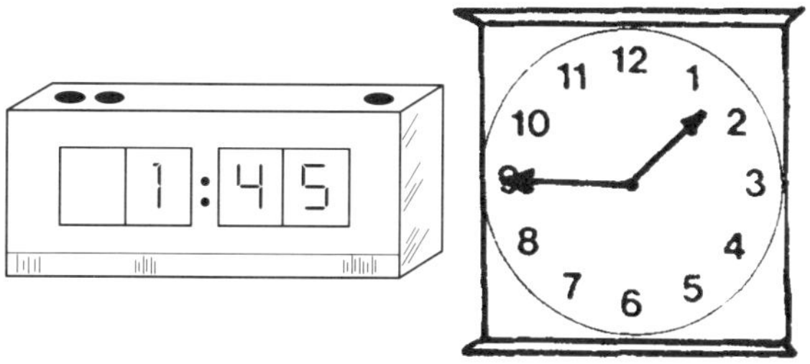
8:15 eight-fifteen	1:45 one forty-five

What time is it?

1. 

   _____ : _____

2. 

   _____ : _____

3. 

   _____ : _____

4. 

   _____ : _____

5. 

   _____ : _____

6. 

   _____ : _____

Draw the hands to show the time.

7. 

   9:30

8. 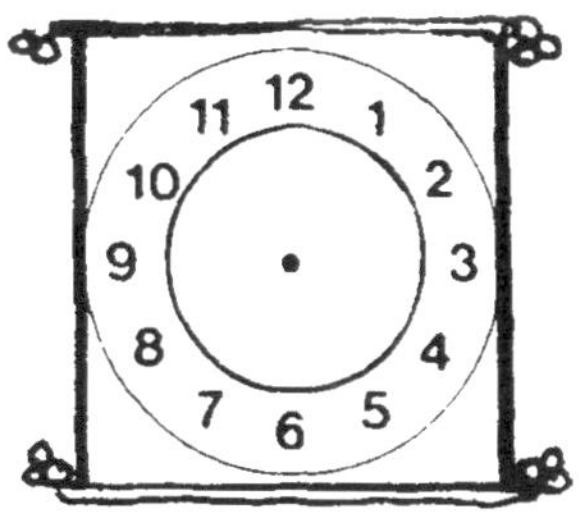

   7:15

9. 

   5:45

Elapsed Time

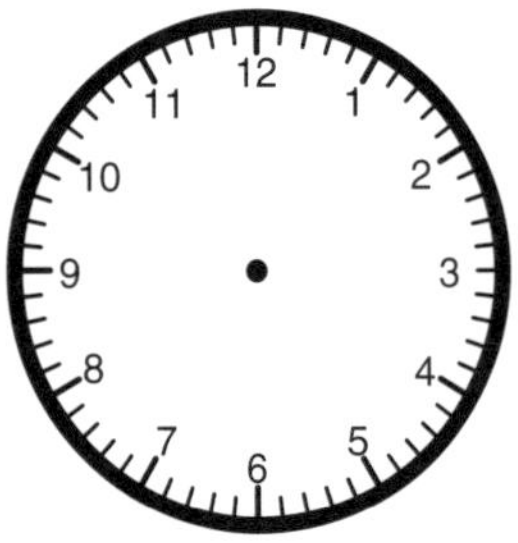

start school

______ : ______

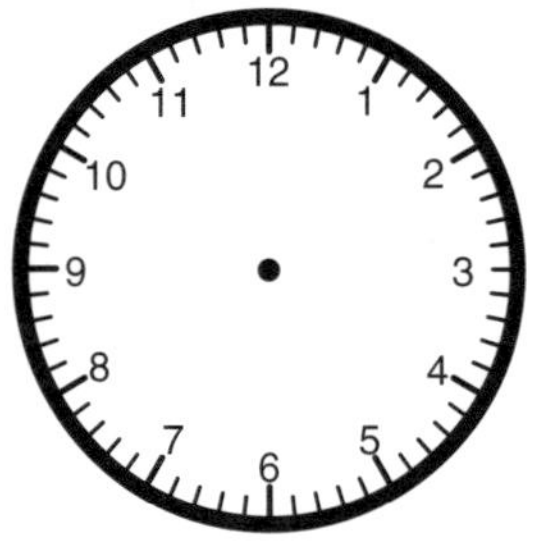

eat lunch

______ : ______

Show the hands of the clock at 9:00 o'clock and 12:00 o'clock.

How much time was spent at school in the morning? ______hours

Write the time under each clock.

---

1. Write both times. Write how many hours later.

______ : ______

______ : ______ _______ hours later

2.

______ : ______

______ : ______ _______ hours later

---

3. What time will this job end?

Job	Start	Amount of Time
Walk to Store	4:30	10 minutes

______ : ______

4. What time will this job end?

Job	Start	Amount of Time
Clean Room	9:00	20 minutes

______ : ______

Measuring Lines

Write the answer above the line.

How long is each line to the nearest $\frac{1}{2}$ inch?

1.

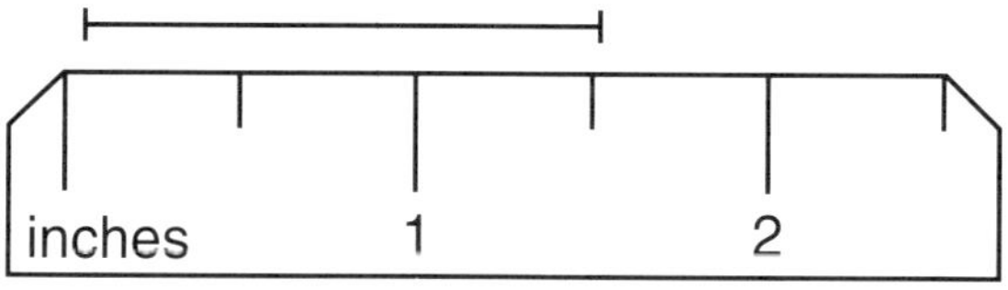

2.

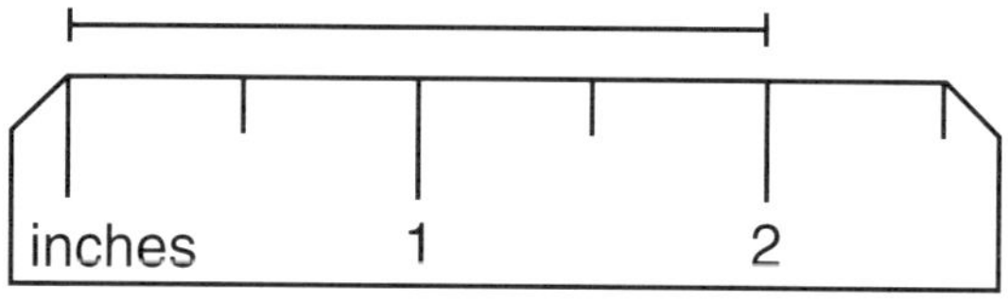

3.

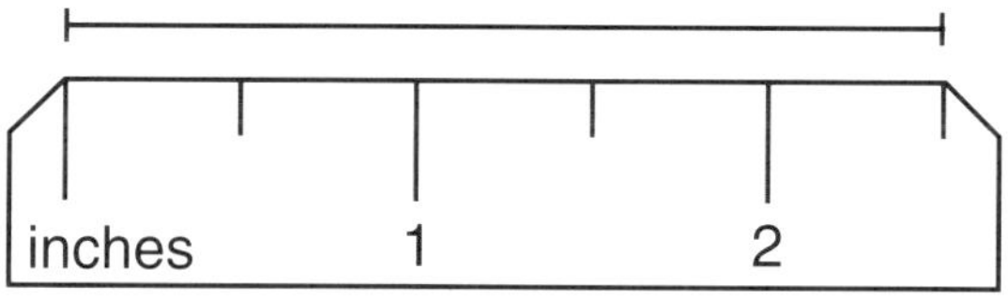

4.

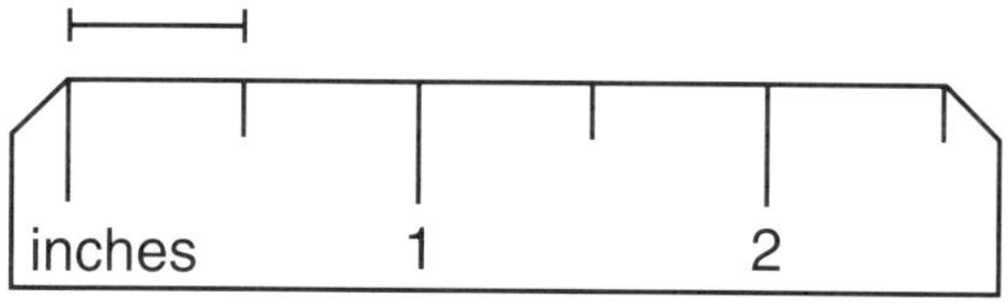

5.

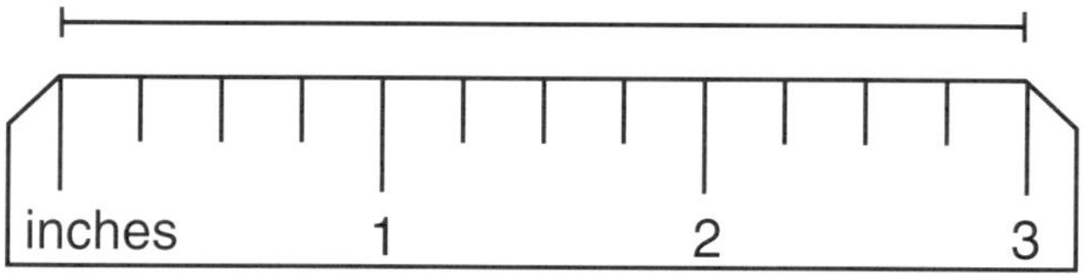

6.

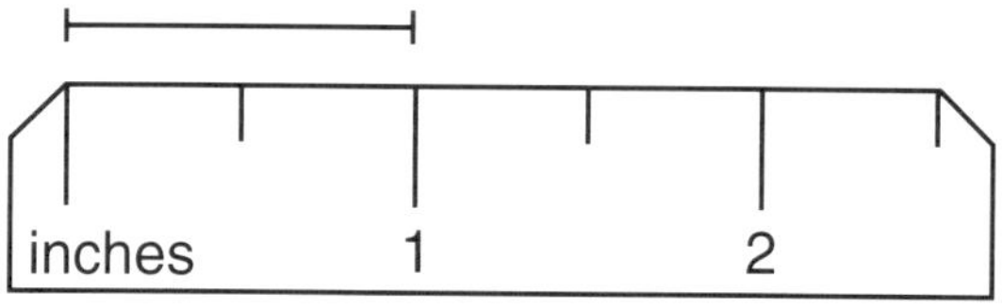

7.

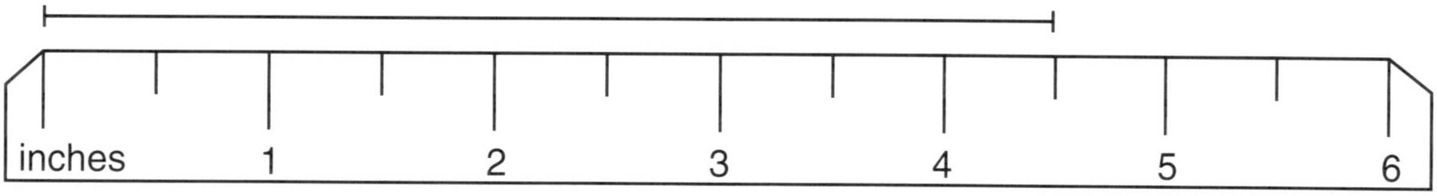

8.

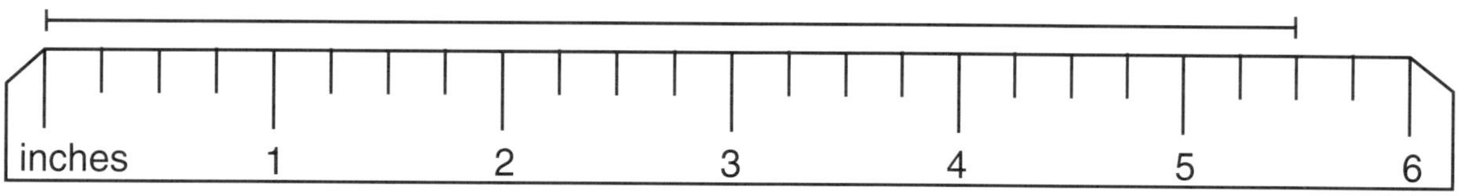

Customary Units of Length

(not actual size)

12 inches = 1 foot

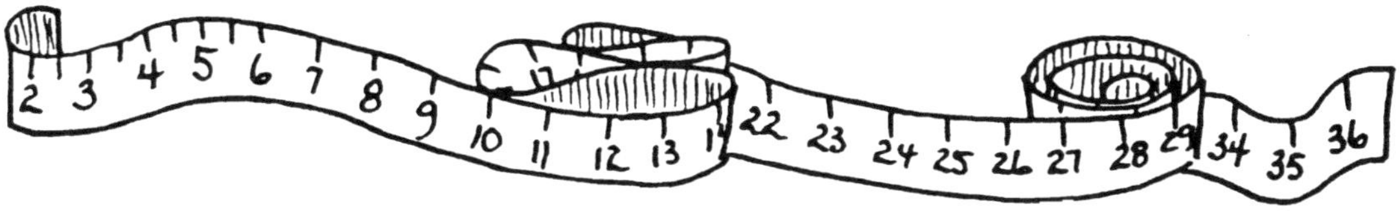

36 inches = 1 yard (yd.)
3 feet (ft.) = 1 yard

Complete.

1. 1 ft. = _____ in. **2.** 2 ft. = _____ in. **3.** 3 ft. = _____ in.

4. 12 in. = _____ ft. **5.** 36 in. = _____ ft. **6.** 24 in. = _____ ft.

7. 1 yd. = _____ ft. **8.** 2 yd. = _____ ft. **9.** 5 yd. = _____ ft.

10. 12 ft. = _____ yd. **11.** 9 ft. = _____ yd. **12.** 24 ft. = _____ yd.

13. 36 in. = _____ yd. **14.** 2 yd. = _____ in. **15.** 5 yd. = _____ in.

16. 1 ft. 2 in. = _____ in. **17.** 2 ft. 6 in. = _____ in.

18. 15 in. = _____ ft. _____ in. **19.** 26 in. = _____ ft. _____ in.

Millimeters, Centimeters and Meters

The centimeters on this ruler have been divided into 10 smaller, equal parts. Each small part or space is 1 millimeter.

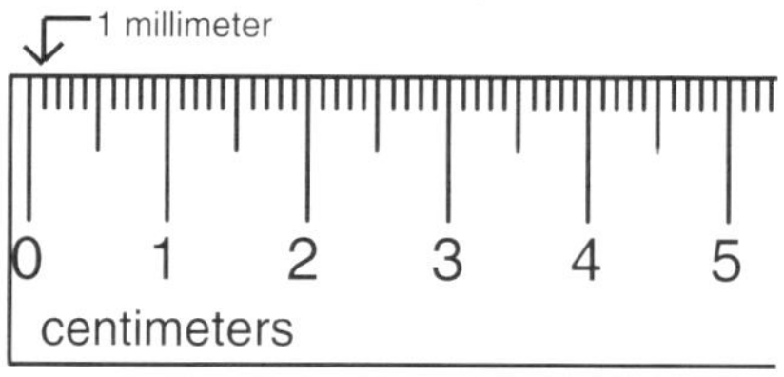

1 centimeter (cm) = 10 millimeters (mm)

A long centimeter ruler with 100 centimeters on it is called a meter stick.

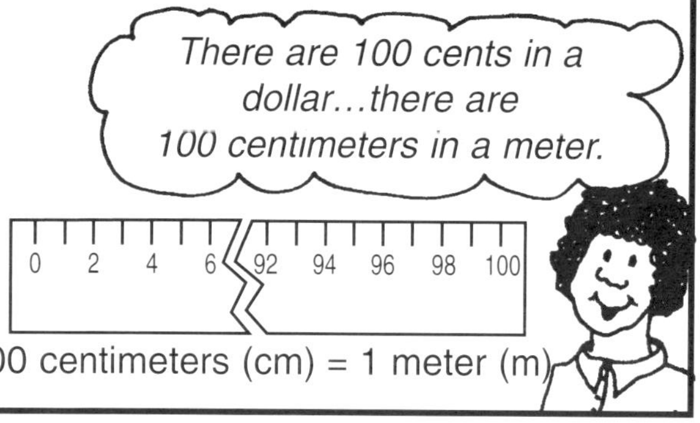

100 centimeters (cm) = 1 meter (m)

1. 1 cm = _____ mm **2.** 2 cm = _____ mm **3.** 4 cm = _____ mm

4. 6 cm = _____ mm **5.** 10 cm = _____ mm **6.** 100 cm = _____ mm

7. 1 m = _____ cm **8.** 2 m = _____ cm **9.** 4 m = _____ cm

10. 10 mm = _____ cm **11.** 20 mm = _____ cm **12.** 100 mm = _____ cm

13. 100 cm = _____ m **14.** 200 cm = _____ m **15.** 500 cm = _____ m

Millimeter, centimeter or meter?

16. thickness of a dime __________ **17.** thickness of a period __________

18. height of a desk __________ **19.** width of little finger __________

How tall are you in centimeters?
How tall are you in millimeters?
About how tall are you in meters?

Finding Perimeter

The distance around the outside of a figure is its **perimeter**.
One way to find perimeter is to lay units or cubes along the outside edges and count the cubes.

Count and number the cubes along the outside edges.
The perimeter of the rectangle is 14 centimeters.

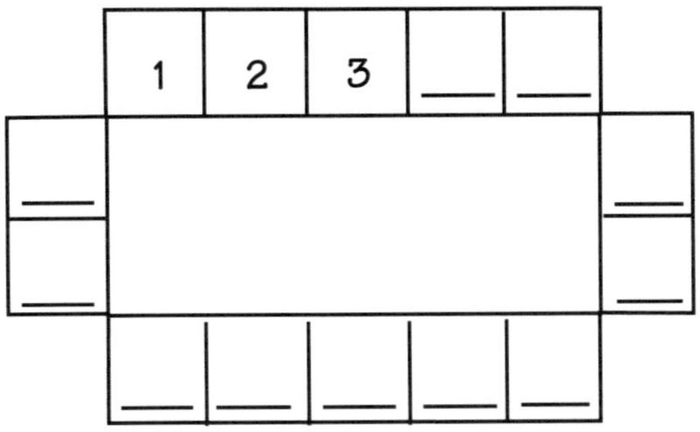

About how many centimeter blocks fit around the perimeter of each figure?

1.

Perimeter = ________ cm

2.

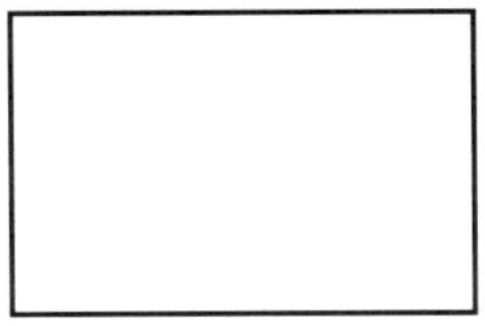

Perimeter = ________ cm

3.

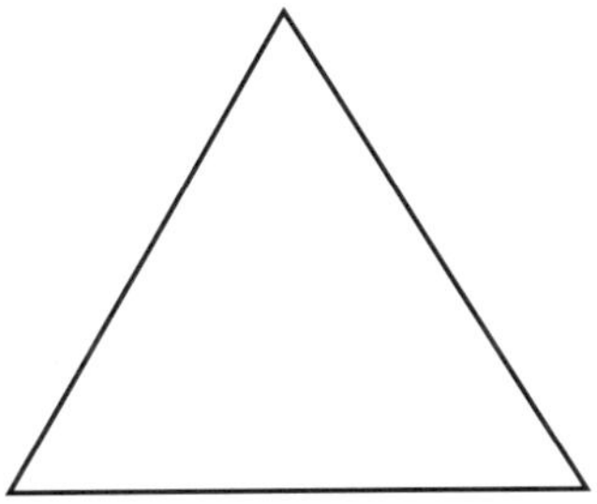

Perimeter = ________ cm

4.

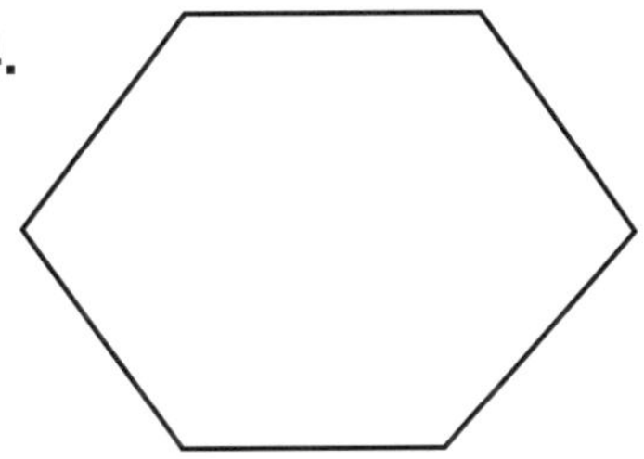

Perimeter = ________ cm

5.

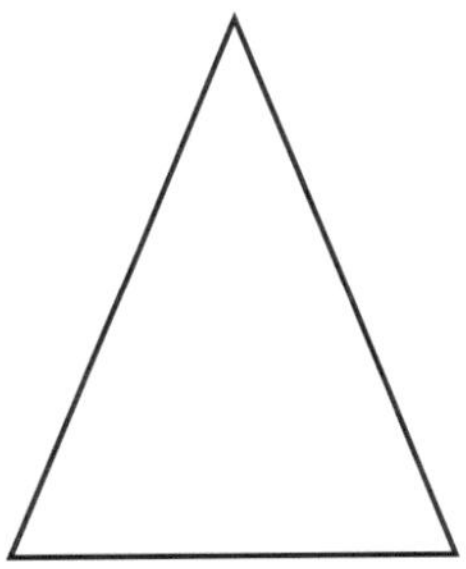

Perimeter = ________ cm

6.

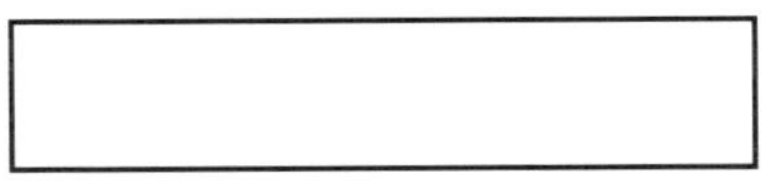

Perimeter = ________ cm

Find the perimeter of each figure.

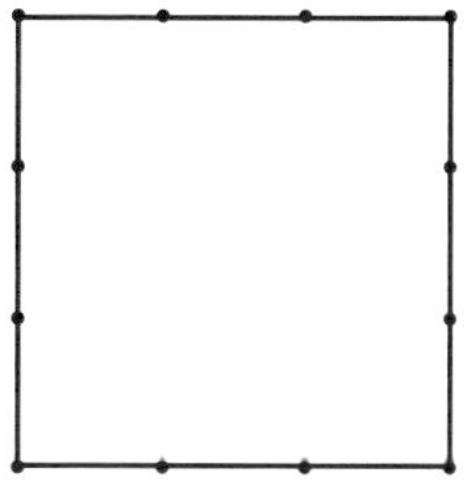

1. perimeter = __________

2. perimeter = __________

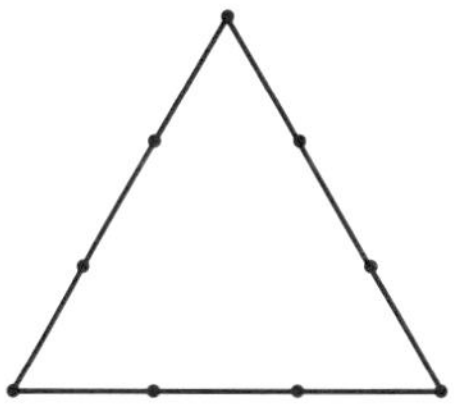

3. perimeter = __________

4. perimeter = __________

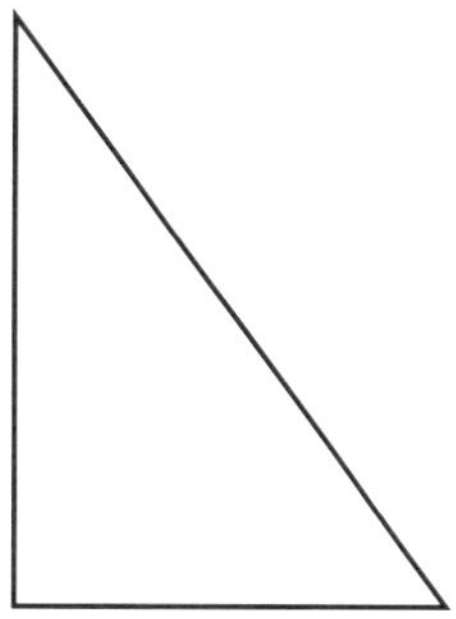

5. perimeter = __________

6. perimeter = __________

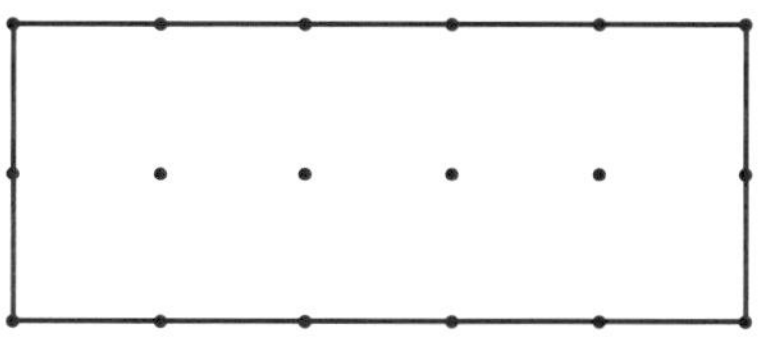

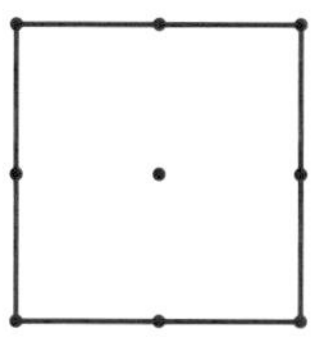

7. perimeter = __________

8. perimeter = __________

How many different ways do you know to find the perimeter?

Area

The number of square units needed to cover the surface of a figure is its **area**.

Count the number of patches on the quilt.

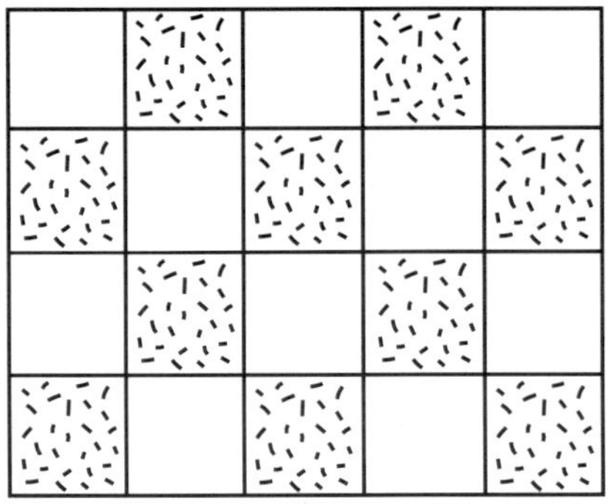

The area is ________ patches

or ________ square units.

1. The diagram shows a square trivet made of tiles. What is the area of the trivet?

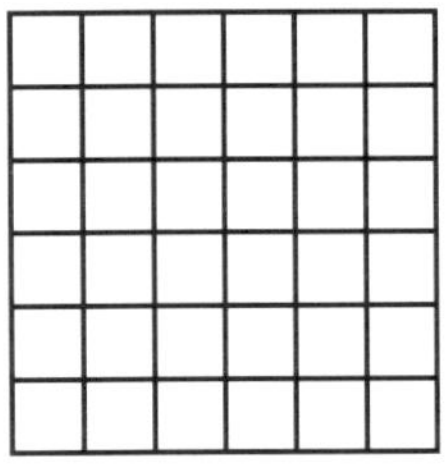

________ tiles

2. The diagram shows a rectangular trivet made of tiles. What is the area of the trivet?

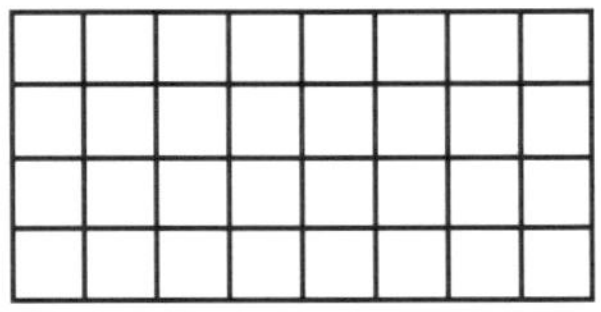

________ tiles

3. Place mats are made by pasting small squares of different colors on construction paper. Each small square is 1 square inch. What is the area of the place mat?

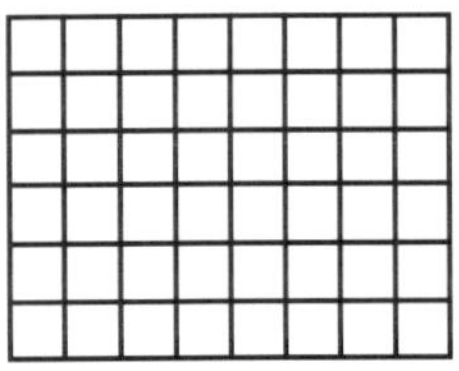

________ square inches

4. A checker board is made of small square units. What is the area of the checker board?

________ square units

5. The diagram shows a kitchen floor. Each tile is 1 square foot. What is the area of the kitchen?

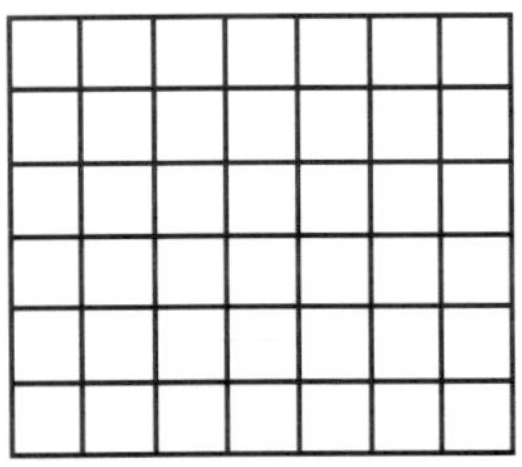

________ square feet

6. The diagram shows an entry way floor. Each tile is 1 square foot. What is the area of the entry?

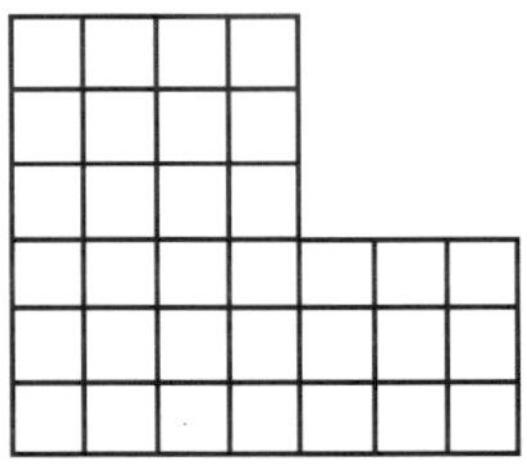

________ square feet

Making a Graph

Color a box for each tally mark.

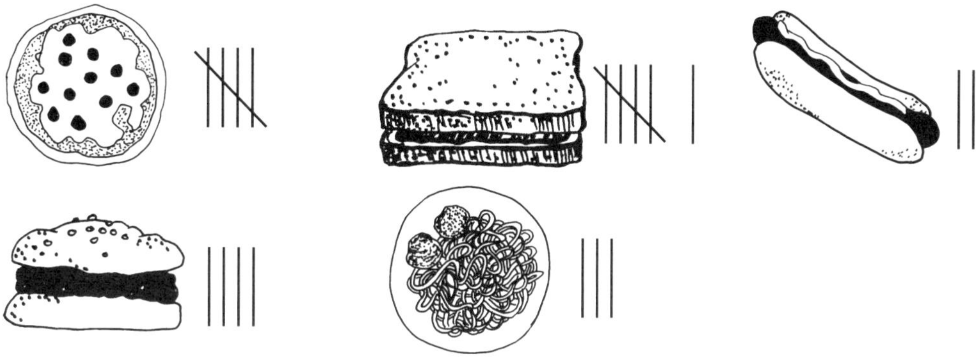

Favorite Lunches

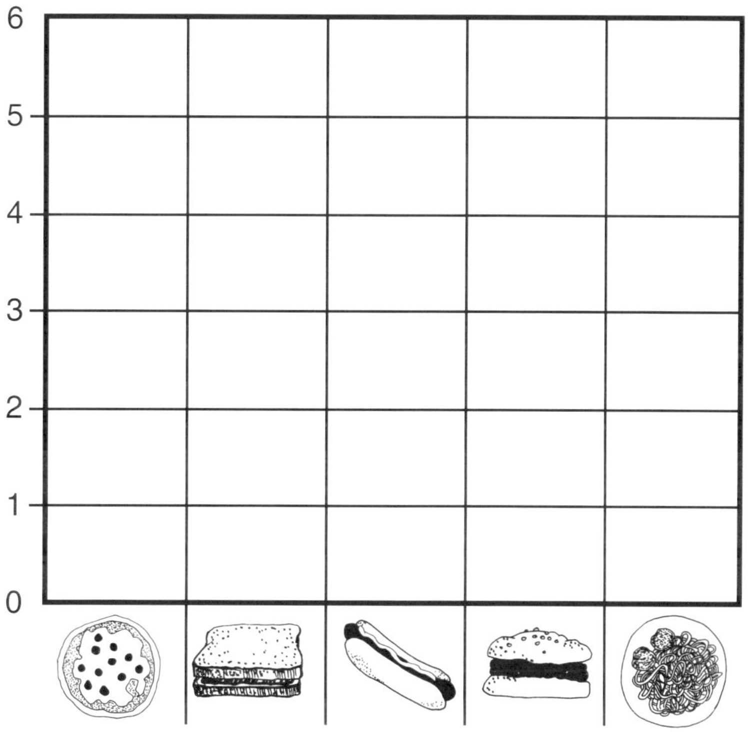

How many chose 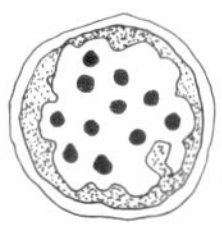 ? ______ 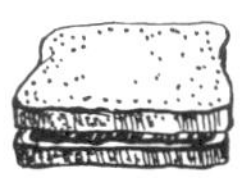 ? ______

 ? ______ 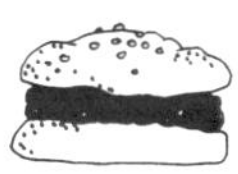 ? ______  ? ______

What is the most popular lunch? ____________________________

Reading Graphs

Study graphs carefully. Read the title and all the words at the bottom and side of graphs.

This graph has a key telling you how much each picture symbol stands for.
Find the key and explain it in your own words.

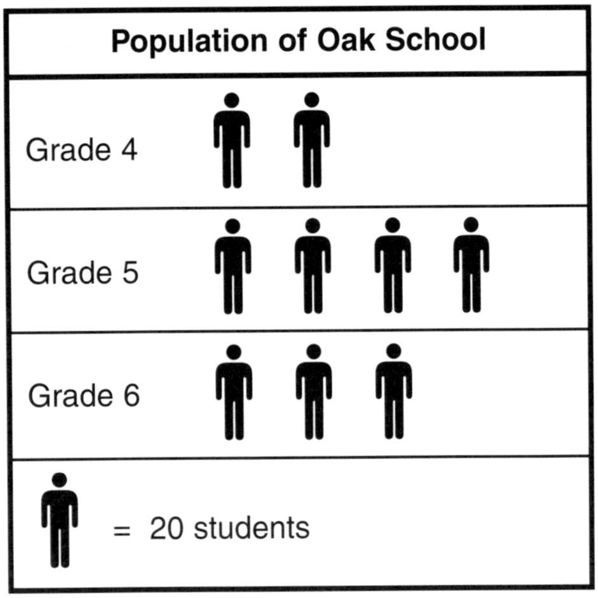

1. How many students in grade 4? ________

2. How many students in grade 5? ________

3. How many students in grade 4 and grade 5 combined? ________

4. How many more students in grade 5 than in grade 4? ________

Average Weekly Allowances	
Grade 4	$ $ $
Grade 5	$ $ $ $
Grade 6	$ $ $ $ $ $
$ stands for $1.00	

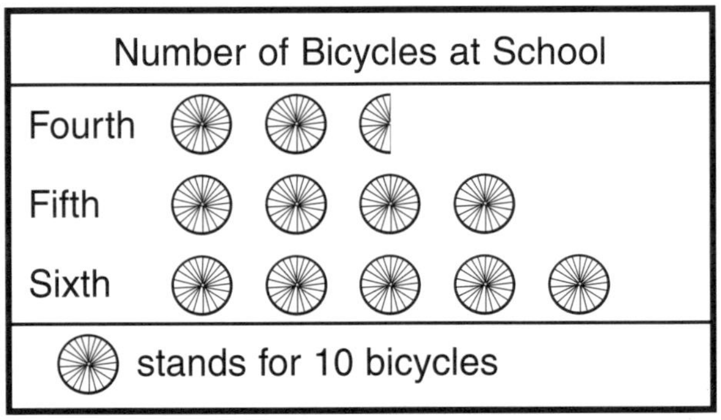

5. What is the average weekly allowance for fourth graders? ________

6. What is the average weekly allowance for a fifth grader? ________

7. How much more allowance does the average fifth grader receive than the average fourth grader? ________

8. How many fourth graders ride bikes to school? ________

9. How many more bikes do sixth graders ride than fifth graders? ________

10. How many fourth, fifth, and sixth graders ride bikes? ________

1. Order these numbers from least to greatest:

   1345 1450 1350 1400 1453

   _____ _____ _____ _____ _____

2. Write as a numeral: __________

   twenty-four thousand four hundred seventy-six

3. Write 502,400 in words:

   ____

   ____

   ____

4. What digit is in the hundred thousands place in 523,467? __________

5. Round 5648 to the nearest ten: __________

Daily Review 1

1. Complete the patterns:

   3, 6, ____, ____, ____, 18

2. Round 5327 to the nearest hundred: __________

3. What number goes in the box to make the sentence true? __________

   $(5 + 2) + 4 = 5 + (\square + 4)$

4. $$\begin{array}{r} 24 \\ 16 \\ 12 \\ +\ 58 \\ \hline \end{array}$$

5. $$\begin{array}{r} 5796 \\ +\ 8057 \\ \hline \end{array}$$

Daily Review 2

1.

$$\begin{array}{r} 25 \\ 12{,}574 \\ 7{,}303 \\ 6 \\ +\quad 125 \\ \hline \end{array}$$

2. What is the name of the answer in an addition problem?

____

3.

$$\begin{array}{r} 504 \\ -\ 245 \\ \hline \end{array}$$

4.

$$\begin{array}{r} 5721 \\ -\ \ 139 \\ \hline \end{array}$$

5. 6325 – 4123 = __________

Daily Review 3

1. 243 + 3124 + 7 = __________

2. In the problem,
6 + 5 = 11,
11 is called the _______________.

3.

$$\begin{array}{r} 305 \\ -\ 124 \\ \hline \end{array}$$

4.

$$\begin{array}{r} 9617 \\ -\ 5248 \\ \hline \end{array}$$

5. 4372 – 251 = __________

Daily Review 4

1. $\begin{array}{r} 603 \\ -\ 249 \\ \hline \end{array}$

2. $\begin{array}{r} 7143 \\ -\ 3238 \\ \hline \end{array}$

3. 6352 – 47 = ________

4. In the problem
9 – 5 = 4,
the 4 is called the ____________.

5. $\begin{array}{r} 8 \\ \times\ 6 \\ \hline \end{array}$ $\qquad$ $\begin{array}{r} 7 \\ \times\ 7 \\ \hline \end{array}$

Daily Review 5

1. Which of these numbers is the greatest? ________

15,271	17,521	15,712	17,251

2. Write as a numeral: ____________

three hundred sixty-two thousand seven hundred

3. Write 480,912 in words:

____

____

____

4. What is the name of the place value of the 4 in 548,631?

____

5. Round 6534 to the nearest ten.

____

Daily Review 6

Daily Reviews 7 and 8

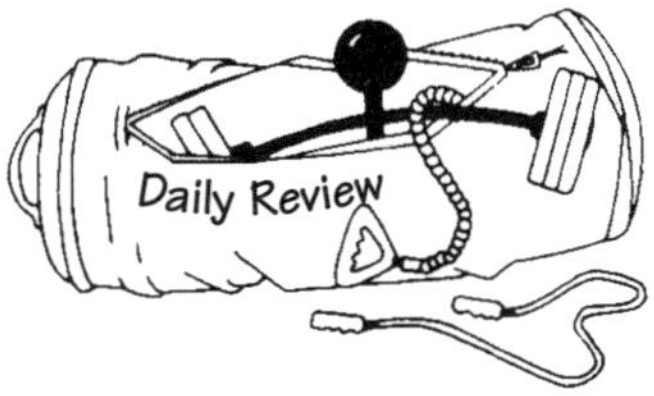

1. $\begin{array}{r} 619 \\ \times \quad 4 \\ \hline \end{array}$

2. $\begin{array}{r} 30 \\ \times \quad 2 \\ \hline \end{array}$ $4 \times 20 =$ ______

3. What is the name of the answer in a multiplication problem?

____

4. $8\overline{)64}$ $72 \div 8 =$ ______

5. $3\overline{)43}$

Daily Review 7

1. Complete the pattern:

7, 14, _____, _____, _____, 42

2. Round 3518 to the nearest hundred: __________

3. What number goes in the box to make the sentence true? _________

$(4 + 2) + 8 = 4 + (\square + 8)$

4. $\begin{array}{r} 42{,}359 \\ +\ 35{,}444 \\ \hline \end{array}$

5. $5\overline{)60}$

Daily Review 8

1. $$\begin{array}{r} 704 \\ -\ 269 \\ \hline \end{array}$$

2. $$\begin{array}{r} 42{,}913 \\ -\ 20{,}342 \\ \hline \end{array}$$

3. $53{,}084 - 6{,}452 =$ __________

4. What is the answer in a subtraction problem called?

____

5. $4\overline{)61}$

Daily Review 9

1. $5326 + 7 + 423 =$ __________

2. $9 \times 7 =$ _______ $$\begin{array}{r} 8 \\ \times\ 8 \\ \hline \end{array}$$

3. $$\begin{array}{r} 508 \\ \times\quad 3 \\ \hline \end{array}$$

4. $$\begin{array}{r} 43 \\ \times\ 60 \\ \hline \end{array}$$

5. What is the answer in a multiplication problem called?

____

Daily Review 10

Daily Reviews 11 and 12

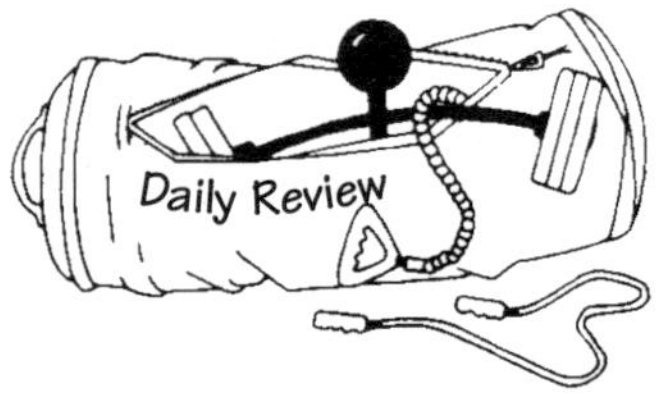

1. Order from least to greatest:

1001   999   1002   1000   998

____ ____ ____ ____ ____

2. Complete the pattern:

9, 18, 27, _____, _____, 54

3. Write 502,416 in words:

____

____

____

4. Round 7625 to the nearest ten: ________

5. $1204 + 56 + 351 + 8 =$ ________

Daily Review 11

1. Round 6573 to the nearest hundred: ________

2. What number goes in the box? ________

$(3 + 5) + 4 = \square + (5 + 4)$

3. $9\overline{)63}$ $54 \div 6 =$ ________

4. $6\overline{)72}$

5. Which shows the quotient of 12 and 3? ________

A $12 + 3$ **C** 12×3

B $12 - 3$ **D** $12 \div 3$

Daily Review 12

1. $\begin{array}{r} 9 \\ \times\ 9 \\ \hline \end{array}$ $7 \times 7 =$ ______

2. $\begin{array}{r} 708 \\ \times\ 8 \\ \hline \end{array}$

3. $\begin{array}{r} 63 \\ \times\ 20 \\ \hline \end{array}$

4. $9\overline{)81}$ $6\overline{)54}$

5. $7\overline{)84}$

Daily Review 13

1. $\begin{array}{r} 503 \\ -\ 142 \\ \hline \end{array}$

2. $\begin{array}{r} 8930 \\ -\ 5245 \\ \hline \end{array}$

3. $7531 - 248 =$ ______

4. $8\overline{)64}$ $56 \div 7 =$ ______

5. $8\overline{)93}$

Daily Review 14

1. What time is it? ________

2. What is the temperature? ________

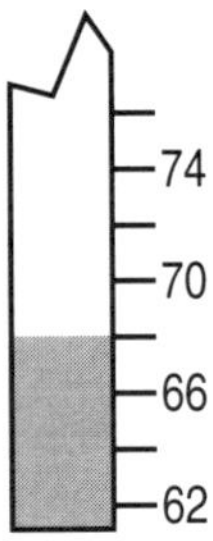

3. How long is the line to the nearest .5 cm? ________

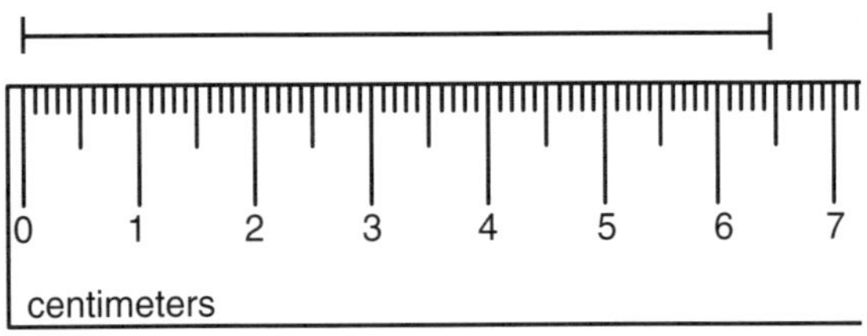

4. Tom typed 24 words per minute for 3 minutes. How many words did he type? ________

5. There is room for 4 students at each table in the library. How many tables will 36 students fill? ________

Daily Review 15

1. $\begin{array}{r} 36 \\ \times\ 21 \\ \hline \end{array}$

2. **12 inches = 1 foot**
3 feet = 1 yard

2 feet = ________ inches

3. **1 meter = 100 centimeters**

4 meters = ________ centimeters

4.

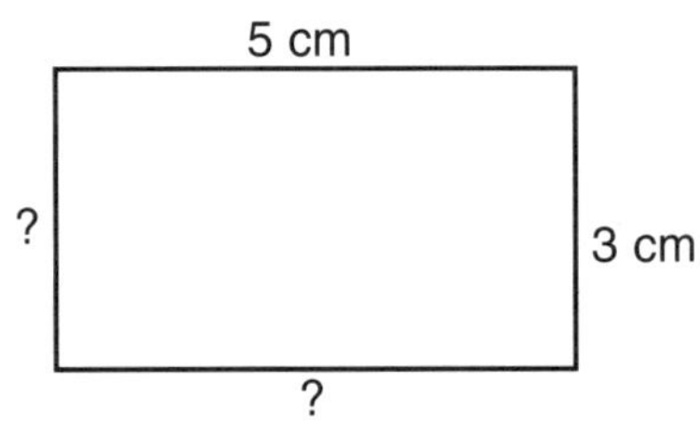

How long are the missing sides?

________ cm and ________ cm

5. You bought a ruler for $0.69. You gave the clerk a $1.00 bill. How much change should you receive? ________

Daily Review 16

1. $3\overline{)9367}$

2. $2\overline{)4602}$

3. Which sign of operation is used to find the quotient of two numbers? ______

A + **C** ×

B − **D** ÷

4. Which is a line segment? ______

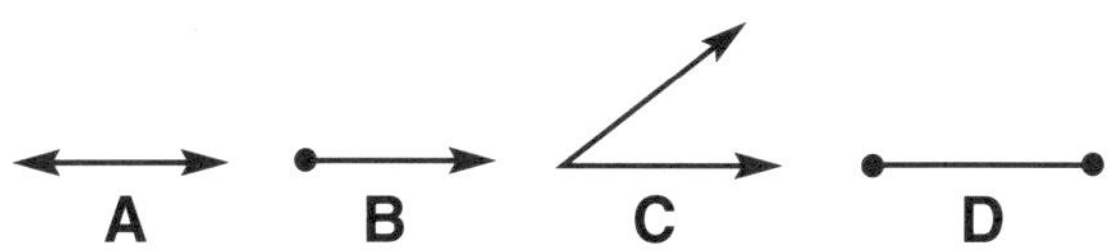

5. A slanting line is a diagonal line. Which line is diagonal? ______

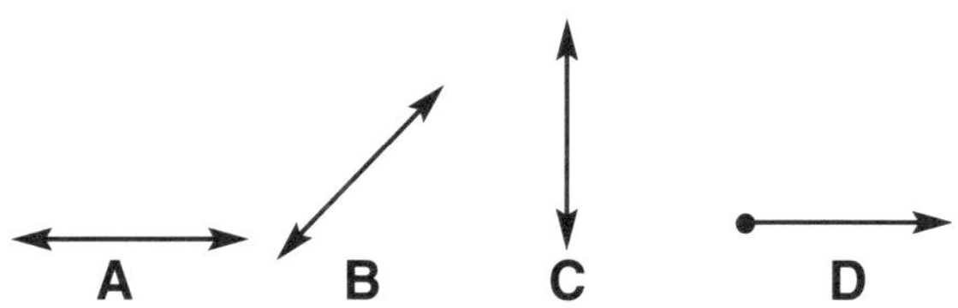

1. What time is it? ______

2. How long is the line to the nearest .5 centimeter? ______

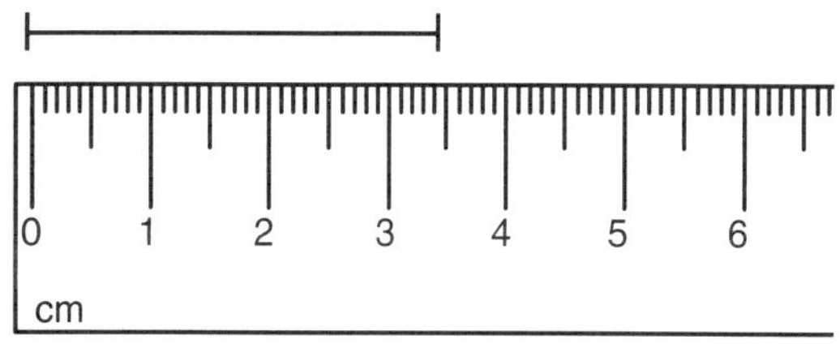

3. You buy a lunch for $3.65. How much change should you receive from a $5.00 bill? ______

4. Jackie bought 3 pens for $0.95 each. How much did Jackie spend for the pens? ______

5. Seven friends wanted to share 42 pieces of pizza equally. How many pieces of pizza should each person get? ______

1. What fractional part is shaded? ________

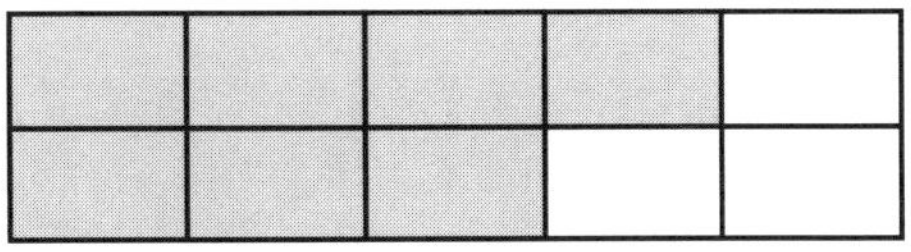

2. What other name for 1 is shown? ________

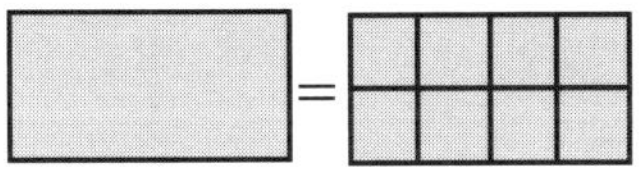

1 = ________

3. Shade in the correct answer. Fill in the blank.

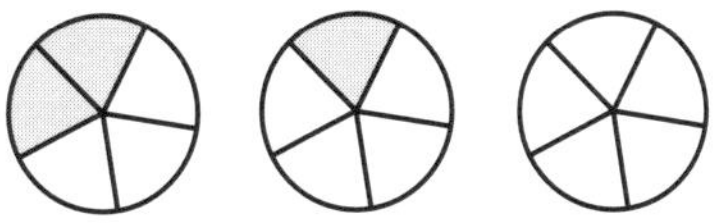

$\frac{2}{5} + \frac{1}{5} =$ ________

4. $\begin{array}{r} 5\frac{3}{5} \\ -\ 2\frac{1}{5} \\ \hline \end{array}$

5. Which lines are parallel? ________

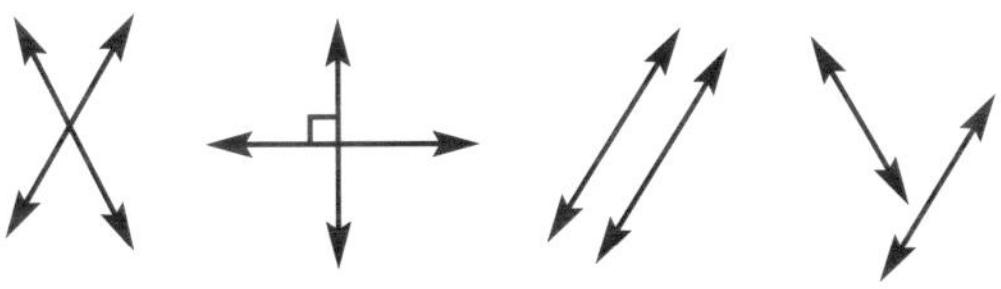

Daily Review 19

1. What fraction of the set is shaded? ________

2. Circle other names for 1:

$\frac{2}{2}$ $\frac{3}{4}$ $\frac{1}{5}$ $\frac{6}{6}$ $\frac{8}{8}$

3. $\begin{array}{r} \frac{7}{8} \\ -\ \frac{4}{8} \\ \hline \end{array}$ $\frac{4}{7} + \frac{2}{7} =$ ________

4. $\begin{array}{r} 3\frac{3}{7} \\ +\ 1\frac{2}{7} \\ \hline \end{array}$

5. Which lines are parallel? ________

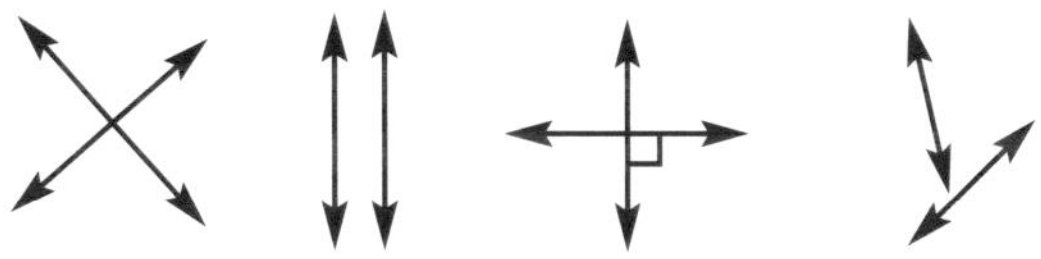

Daily Review 20

1.

2 cups = 1 pint
2 pints = 1 quart

2 quarts = _______ cups

2.

100 centimeters = 1 meter
1000 meters = 1 kilometer

600 centimeters = _______ meters

3. What is the perimeter of this figure? ________ in.

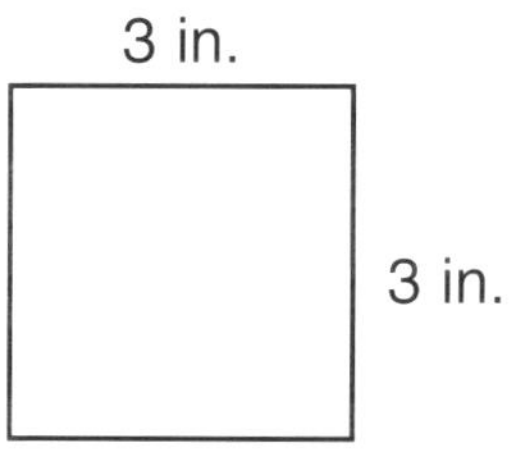

4. David sold 9 glasses of lemonade at 5¢ per glass. How much money did he collect?

____

5. Four friends wish to share 40 cards equally. How many cards will each person get?

____

1. $$\begin{array}{r} 52 \\ \times\ 37 \\ \hline \end{array}$$

2. $4\overline{)172}$

3. $2\overline{)6407}$

4. What is the name of the figure?

____

5. Which line is horizontal? ________

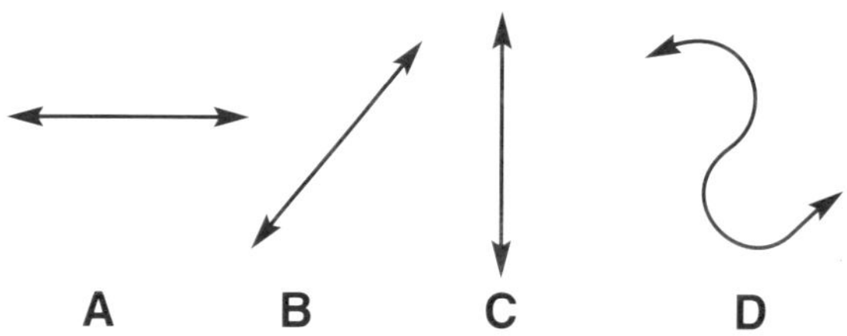

Daily Reviews 23 and 24

1. In the number 472, the 7 is in what place? __________

2. Order from least to greatest:

   7762   3276   3270   3207   3700

   ____ ____ ____ ____ ____

3. What number goes in the box? ________

   8, 16, □, 32, 40

4. Write as a numeral: __________

eight thousand three hundred fifty-six

5. Write 23,508 in words:

   ____

   ____

   ____

Daily Review 23

1. What is the name of the place value of the 6 in 465,237?

   ____

2. Round 642 to the nearest ten: __________

3. Round 3674 to the nearest hundred: __________

4. What number goes in the box to make this sentence true? ________

   $(6 + 3) + 7 = 6 + (\square + 7)$

5. $$\begin{array}{r} 493 \\ +\ 258 \\ \hline \end{array}$$

Daily Review 24

1.
$$\begin{array}{r} 43 \\ 25 \\ 80 \\ +\ 14 \\ \hline \end{array}$$

2.
$$\begin{array}{r} 36{,}537 \\ +\ 47{,}124 \\ \hline \end{array}$$

3. $6 + 507 + 3241 + 52 =$ ________

4. What is 13 called in the problem $8 + 5 = 13$?

____

5.
$$\begin{array}{r} 762 \\ -\ 184 \\ \hline \end{array}$$

Daily Review 25

1.
$$\begin{array}{r} 802 \\ -\ 247 \\ \hline \end{array}$$

2.
$$\begin{array}{r} 6465 \\ -\ 2817 \\ \hline \end{array}$$

3. $53{,}426 - 253 =$ ________

4. Which shows the difference between 10 and 5? ________

A $10 + 5$ **C** 10×5

B $10 - 5$ **D** $10 \div 5$

5. $9 \times 6 =$ ________
$$\begin{array}{r} 8 \\ \times\ 7 \\ \hline \end{array}$$

Daily Review 26

Daily Reviews 27 and 28

1. $\begin{array}{r} 467 \\ \times \quad 8 \\ \hline \end{array}$

2. $\begin{array}{r} 65 \\ \times \quad 30 \\ \hline \end{array}$

3. $\begin{array}{r} 74 \\ \times \quad 29 \\ \hline \end{array}$

4. What is the answer in a multiplication problem called?

   ____

5. $6\overline{)48}$ $63 \div 7 =$ ________

Daily Review 27

1. $6\overline{)74}$

2. $3\overline{)642}$

3. $4\overline{)8092}$

4. Which shows how to find the quotient of 10 and 5? ________

   A $10 + 5$ **C** 10×5

   B $10 - 5$ **D** $10 \div 5$

5. What fractional part is shaded? ________

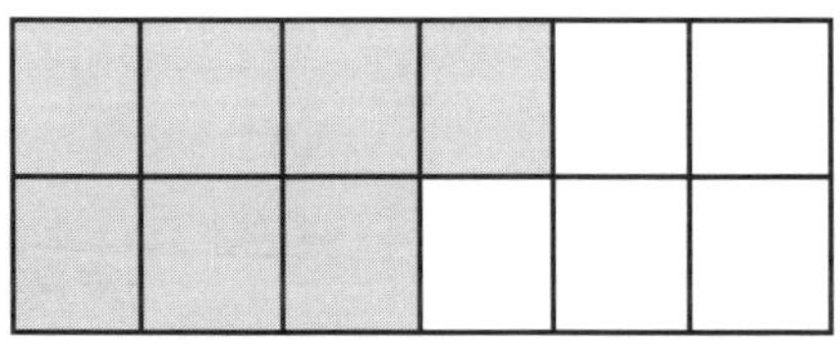

Daily Review 28

1. What fraction of the set is circled? ________

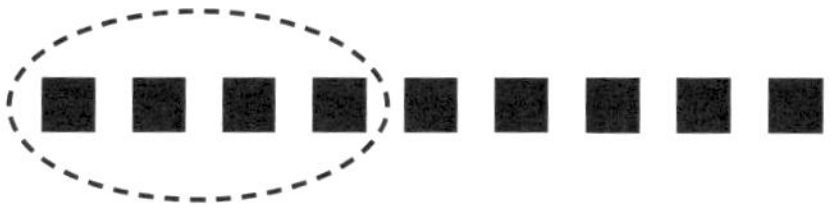

2. Which symbol (>, =, <) goes in the box? ________

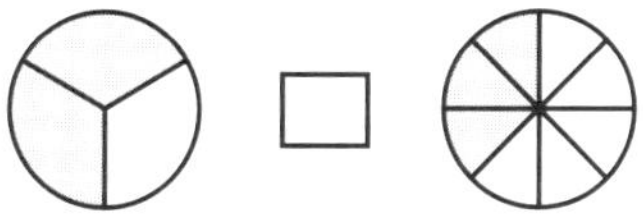

3. $\frac{4}{5} - \frac{1}{5} =$ ________

$$\begin{array}{r} \frac{4}{9} \\ + \ \frac{3}{9} \\ \hline \end{array}$$

4. $$\begin{array}{r} 4\frac{3}{8} \\ + \ 5\frac{2}{8} \\ \hline \end{array}$$

5. What is the name of this figure?

____

Daily Review 29

1. Which line is vertical? ________

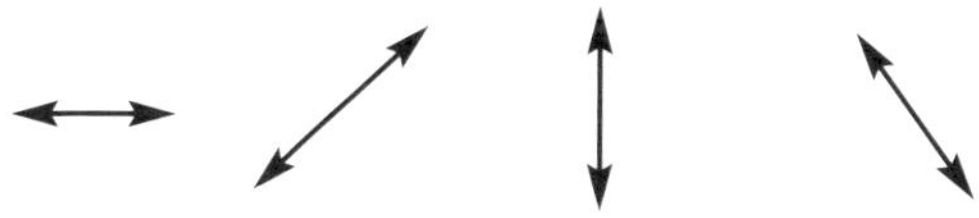

A B C D

2. Which lines are parallel? ________

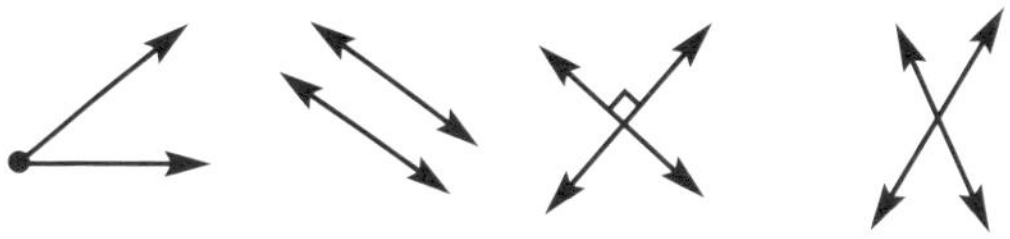

A B C D

3. Which shows a line of symmetry? ________

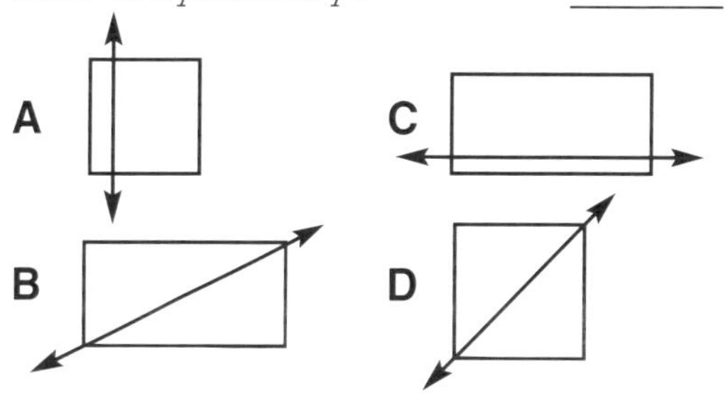

4. Which figures are congruent? ________

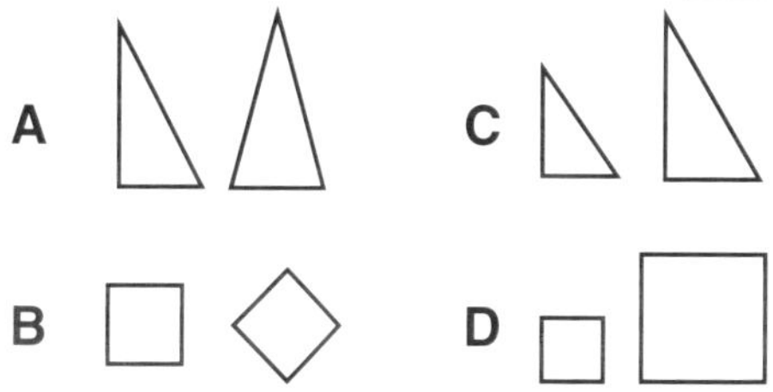

5. What is the name of this figure?

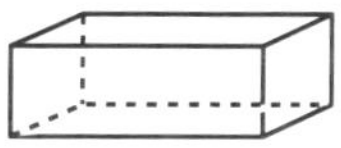

____

A cylinder
B sphere
C pyramid
D rectangular solid

Daily Review 30

1. What time is it?

____

2. What is the temperature?

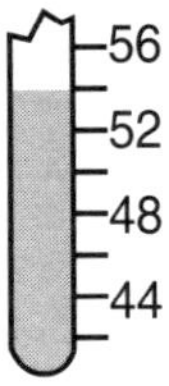

____

3. What is the measure of the line to the nearest half-inch? __________

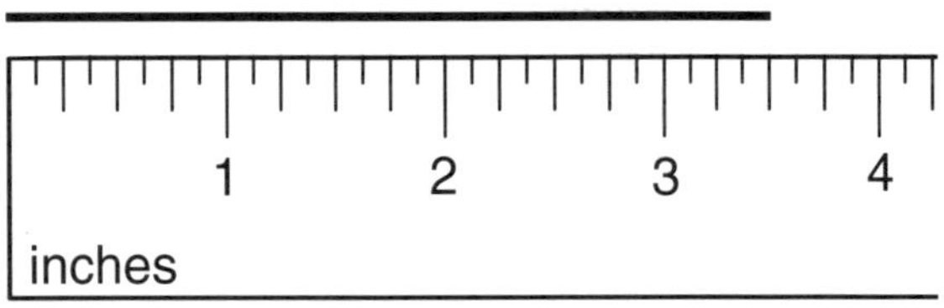

4.

16 ounces = 1 pound
2000 pounds = 1 ton

3 tons = ________ pounds

5.

100 centimeters = 1 meter
1000 meters = 1 kilometer

5000 meters = _______ kilometers

Daily Review 31

1. What is the perimeter of this figure? _______ inches

5 inches

2. You bought a hat for $2.45 and a book for $1.52. How much change should you receive from a $5.00 bill? __________

3. A baseball team has 9 players. How many players are there on 8 baseball teams? __________

4. A group of 54 students are to be put into baseball teams of 9 players each. How many teams will there be? __________

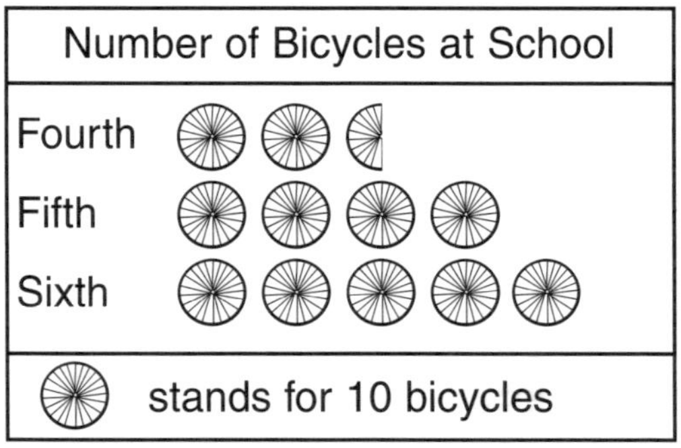

5. How many students in fifth grade have bicycles at school? __________

Daily Review 32